CONTENTS

Make: Volume 66 Dec 2018/Jan 2019

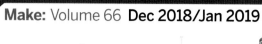

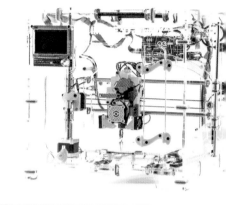

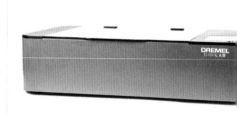

06

52

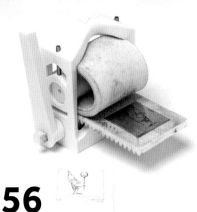

56

ON THE COVER:
The Pulse XE 3D printer, Glowforge laser cutter, and Wazer waterjet cutter wowed in this year's testing.
Photo: Mark Madeo

62

Kelly Egan, Nick Ervinck, Lady Red Beacham, Martin Schneider, Tasker Smith

Make:

> "Technology will not replace great teachers, but technology in the hands of great teachers can be transformational."
> — George Couros

EXECUTIVE CHAIRMAN & CEO
Dale Dougherty
dale@makermedia.com

CFO & COO
Todd Sotkiewicz
todd@makermedia.com

EDITORIAL

EDITORIAL DIRECTOR
Roger Stewart
roger@makermedia.com

EXECUTIVE EDITOR
Mike Senese
mike@makermedia.com

SENIOR EDITORS
Keith Hammond
khammond@makermedia.com
Caleb Kraft
caleb@makermedia.com

EDITOR
Laurie Barton

PRODUCTION MANAGER
Craig Couden

BOOKS EDITOR
Patrick Di Justo

CONTRIBUTING EDITORS
William Gurstelle
Charles Platt
Matt Stultz

CONTRIBUTING WRITERS
Matt Bell, Gareth Branwyn, Lydia Sloan Cline, Matt Dauray, Kelly Egan, Alex Glow, Merve Güngör, Nisan Lerea, Olansky Luciano, Winston Moy, Niq Oltman, Ryan Priore, Jen Schachter, Florian Schäffer, Martin Schneider, Star Simpson, Tasker Smith, Mandy L. Stultz, Sarah Vitak, Matthijs Witsenburg, Chris Yohe

DESIGN, PHOTOGRAPHY & VIDEO

ART DIRECTOR
Juliann Brown

SENIOR VIDEO PRODUCER
Tyler Winegarner

MAKEZINE.COM

WEB/PRODUCT DEVELOPMENT
Rio Roth-Barreiro
Maya Gorton
Pravisti Shrestha
Stephanie Stokes
Travis Stone
Alicia Williams

CONTRIBUTING ARTISTS
Kelly Egan, Mark Madeo, Rob Nance, Hep Svadja

ONLINE CONTRIBUTORS
Christie Bender, Jennifer Blakeslee, Chiara Cecchini, Jon Christian, DC Denison, Gretchen Giles, Jonathan Gleich, Grete Kaulinyte, Pinguino Kolb, Bahar Kumar, Vladimir Kuznetsov, Joel Leonard, Connie Liu, Goli Mohammadi, Luigi Pacheco, Pete Prodoehl, Nathan Pritchett, Robert Ryan-Silva, Jaiveer Singh, Violet Su, Christian Zeh

PARTNERSHIPS & ADVERTISING
makermedia.com/contact-sales or partnerships@makezine.com

SENIOR DIRECTOR OF PARTNERSHIPS & PROGRAMS
Katie D. Kunde

DIRECTOR OF PARTNERSHIPS
Shaun Beall

STRATEGIC PARTNERSHIPS
Cecily Benzon
Brigitte Mullin

DIRECTOR OF MEDIA OPERATIONS
Mara Lincoln

DIGITAL PRODUCT STRATEGY

SENIOR DIRECTOR, CONSUMER EXPERIENCE
Clair Whitmer

MAKER FAIRE

MANAGING DIRECTOR
Sabrina Merlo

WEB PRODUCER
Bill Olson

MAKER SHARE

DIGITAL COMMUNITY PRODUCT MANAGER
Matthew A. Dalton

COMMERCE

PRODUCT MARKETING MANAGER
Ian Wang

OPERATIONS MANAGER
Rob Bullington

PUBLISHED BY

MAKER MEDIA, INC.
Dale Dougherty

Copyright © 2018 Maker Media, Inc. All rights reserved. Reproduction without permission is prohibited. Printed in the USA by Schumann Printers, Inc.

Comments may be sent to:
editor@makezine.com

Visit us online:
makezine.com

Follow us:
🐦 @make @makerfaire @makershed
🔘 google.com/+make
📘 makemagazine
📷 makemagazine
▶️ makemagazine
📺 twitch.tv/make
📌 makemagazine

Manage your account online, including change of address: makezine.com/account
866-289-8847 toll-free in U.S. and Canada
818-487-2037,
5 a.m.–5 p.m., PST
cs@readerservices.makezine.com

STATEMENT OF OWNERSHIP, MANAGEMENT AND CIRCULATION (required by Act of August 12, 1970: Section 3685, Title 39, United States Code). 1. MAKE Magazine 2. (ISSN: 1556-2336) 3. Filing date: 10/1/2018. 4. Issue frequency: Bi Monthly. 5. Number of issues published annually:6. 6. The annual subscription price is 34.95. 7. Complete mailing address of known office of publication: Maker Media, Inc. 1005 Gravenstein Highway North, Sebastopol, CA 95472. Contact person: Kolin Rankin. Telephone: 305-859-0063 8. Complete mailing address of headquarters or general business office of publisher: Maker Media, Inc. 1700 Montgomery Street; Suite 240, San Francisco, CA 94111. 9. Full names and complete mailing addresses of publisher, editor, and managing editor. Publisher, Todd Sotkiewicz, Maker Media, Inc., 1700 Montgomery Street; Suite 240, Editor, Mike Senese, Maker Media, Inc., 1700 Montgomery Street; Suite 240, San Francisco, CA 94111, Managing Editor, N/A, Maker Media, Inc., 1700 Montgomery Street; Suite 240, San Francisco, CA 94111. 10. Owner: Maker Media, Inc.; 1700 Montgomery Street; Suite 240, San Francisco, CA 94111. 11. Known bondholders, mortgages, and other security holders owning or holding 1 percent of more of total amount of bonds, mortgages or other securities: None. 12. Tax status: Has Not Changed During Preceding 12 Months. 13. Publisher title: MAKE Magazine. 14. Issue date for circulation data below: Oct/Nov 2018. 15. The extent and nature of circulation: A. Total number of copies printed (Net press run). Average number of copies each issue during preceding 12 months:104,475. Actual number of copies of single issue published nearest to filing date: 93,146. B. Paid circulation. 1. Mailed outside-county paid subscriptions. Average number of copies each issue during the preceding 12 months: 61,917. Actual number of copies of single issue published nearest to filing date: 58,336. 2. Mailed in-county paid subscriptions. Average number of copies each issue during the preceding 12 months: 0. Actual number of copies of single issue published nearest to filing date: 0. 3. Sales through dealers and carriers, street vendors and counter sales. Average number of copies each issue during the preceding 12 months: 9,787. Actual number of copies of single issue published nearest to filing date: 8,530. 4. Paid distribution through other classes mailed through the USPS. Average number of copies each issue during the preceding 12 months: 0. Actual number of copies of single issue published nearest to filing date: 0. C. Total paid distribution. Average number of copies each issue during preceding 12 months: 71,704. Actual number of copies of single issue published nearest to filing date: 66,866. D. Free or nominal rate distribution (by mail and outside mail). 1. Free or nominal Outside-County. Average number of copies each issue during the preceding 12 months: 661. Number of copies of single issue published nearest to filing date: 616. 2. Free or nominal rate in-county copies. Average number of copies each issue during the preceding 12 months: 0. Number of copies of single issue published nearest to filing date: 0. 3. Free or nominal rate copies mailed at other Classes through the USPS. Average number of copies each issue during preceding 12 months: 0. Number of copies of single issue published nearest to filing date: 0. 4. Free or nominal rate distribution outside the mail. Average number of copies each issue during preceding 12 months: 2,101. Number of copies of single issue published nearest to filing date: 2,046. E. Total free or nominal rate distribution. Average number of copies each issue during preceding 12 months: 2,761. Actual number of copies of single issue published nearest to filing date: 2,662. F. Total free distribution (sum of 15c and 15e). Average number of copies each issue during preceding 12 months: 74,465. Actual number of copies of single issue published nearest to filing date: 69,528. G. Copies not Distributed. Average number of copies each issue during preceding 12 months: 30,011. Actual number of copies of single issue published nearest to filing date: 23,618. H. Total (sum of 15f and 15g). Average number of copies each issue during preceding 12 months: 104,475. Actual number of copies of single issue published nearest to filing date: 93,146. I. Percent paid. Average percent of copies paid for the preceding 12 months: 96.29%. Actual percent of copies paid for the preceding 12 months: 96.17% 16. Electronic Copy Circulation: A. Paid Electronic Copies. Average number of copies each issue during preceding 12 months: 24,257. Actual number of copies of single issue published nearest to filing date: 24,004. B. Total Paid Print Copies (Line 15c) + Paid Electronic Copies (Line 16a). Average number of copies each issue during preceding 12 months: 95,960. Actual number of copies of single issue published nearest to filing date: 90,870. C. Total Print Distribution (Line 15f) + Paid Electronic Copies (Line 16a). Average number of copies each issue during preceding 12 months: 98,721. Actual number of copies of single issue published nearest to filing date: 93,532. D. Percent Paid (Both Print & Electronic Copies) (16b divided by 16c x 100). Average number of copies each issue during preceding 12 months: 97.20%. Actual number of copies of single issue published nearest to filing date: 97.15%. I certify that 50% of all distributed copies (electronic and print) are paid above nominal price: Yes. Report circulation on PS Form 3526-X worksheet 17. Publication of statement of ownership will be printed in the Dec/Jan 2019 issue of the publication. 18. Signature and title of editor, publisher, business manager, or owner: Todd Sotkiewicz, Business Manager. I certify that all information furnished on this form is true and complete. I understand that anyone who furnishes false or misleading information on this form or who omits material or information requested on the form may be subject to criminal sanction and civil actions.

CONTRIBUTORS

What's the next tool, technique, or skill you want to learn?

Star Simpson
San Francisco, CA
(Functional Furniture)

I'm working on my meta-skills at the moment: trying to improve my writing.

Martin Schneider
Hampton Wick, U.K.
(La Petite Press)

In the upcoming months I want to learn more about 3D printing, get more familiar with designing mechanical parts and using different materials, and acquire model-making techniques.

Alex Glow
San Francisco, CA
(Archimedes: AI Robot Owl)

Tough choice!! I want to learn to use D3.js, so I can synthesize and visualize data including EEG and some homemade biosensors.

Mark Madeo
San Francisco, CA
(Cover photograph)

Since so much of photography requires looking at a screen, the next skill I'd like to learn will be all hands: sculpting with clay. The messier the better!

Issue No. 66, Dec 2018/Jan 2019. *Make:* (ISSN 1556-2336) is published bimonthly by Maker Media, Inc. in the months of January, March, May, July, September, and November. Maker Media is located at 1700 Montgomery Street, Suite 240, San Francisco, CA 94111. SUBSCRIPTIONS: Send all subscription requests to *Make:*, P.O. Box 17046, North Hollywood, CA 91615-9588 or subscribe online at makezine.com/offer or via phone at (866) 289-8847 (U.S. and Canada); all other countries call (818) 487-2037. Subscriptions are available for $34.99 for 1 year (6 issues) in the United States; in Canada: $39.99 USD; all other countries: $50.09 USD. Periodicals Postage Paid at San Francisco, CA, and at additional mailing offices. POSTMASTER: Send address changes to *Make:*, P.O. Box 17046, North Hollywood, CA 91615-9588. Canada Post Publications Mail Agreement Number 41129568. CANADA POSTMASTER: Send address changes to: Maker Media, PO Box 456, Niagara Falls, ON L2E 6V2

PRINTED WITH SOY INK

Hooray for
World Maker Faire and
More "Maker School"

Dain Elman
@DainElman

Overwhelmed by so many creative projects, innovative products, and inspiring stories at World Maker Faire @makerfaire #wmfny18 @nysci

Dain Elman, Ann-Louise Davidson and Nadia Naffi

Education Makers
@educationmakers

Inspiring panelists @makerfaire today! #WMFNY18

Joel Telling 3D Printing Nerd, @estefanniegg William Osman, @Jackiecrafts @TheBrokenNerd83

8:16 PM - 22 Sep 2018 from New York Hall of Science Rocket Park

MAKER SCHOOL

I'm a Library/Media Specialist who has subscribed to *Make:* magazine for the past two years on behalf of my students and colleagues. Volume 65 of your magazine has been the most useful and beneficial volume to my stakeholders of any I have yet read. Seeing so many different avenues of the maker movement highlighted in one place was both encouraging and enlightening. I'm fortunate to work with students who have many different learning styles as well as challenges; Making has something to offer all of them.

At times, I have found your magazine somewhat difficult to mine for projects that are accessible and achievable for my students. However, your feature, "Maker School" had so much to offer. Seeing so many ways in which students of all ages are creating and innovating using both traditional and 21st century methods was fantastic. If my students had made a top 10 list of subjects they wanted to explore further in their makerspace, I feel that you would have touched on every one.

If I could request one thing on behalf of subscribers similar to myself, it would be that you make "Maker School" into a regular feature in your magazine. It goes so far beyond simple projects and into the larger scope of the maker movement and it's importance in education. Students need to see people like themselves innovating, creating, and exploring the possibilities that this field has to offer!

–Blaine Henderson, via the web

HAVE SOMETHING TO SAY? We want to hear!
Send us your stories, photos, gripes, and successes to editor@makezine.com

---

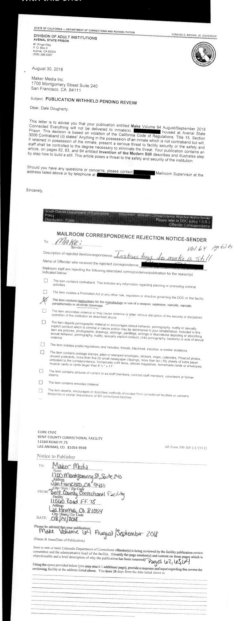

Prison Ban of the Month

» **LOCATION:** Literally everywhere

» **TITLE:** *Make:* Vol. 64, "Invention of the Modern Still"

» **REASON:** "Your publication contains an article, on pages 62, 63, and 64 entitled 'Invention of the Modern Still' describes and illustrates step by step how to build a still."

» *EDITOR'S NOTE: I mean, yeah. That's exactly what we did. Can't really argue with this one.*

UNKNOWN ORIGINS

NICKERVINCK.COM/EN

Artists who also have a deep interest in architecture and engineering frequently approach their work in a way that clearly expresses this cross-disciplinary mindset. Such is the case with Belgian artist **Nick Ervinck**. His 3D printed sculptures are often flowing, complex, interlacing forms that feel both familiar, organic, but also like cyborganic objects from some strange alternative world. In exhibiting his work, Ervinck also considers the space around the piece in ways that increase their impact and drama.

While Ervinck widely exhibits his work as fine art, he also does not shy away from the commercial market. He is always ready to sell his pieces and to accept commissions. For instance, he has used his art and 3D printing skills to design more worldly objects like statues for competition trophies.

Ervinck likes using the technologies of 3D design and printing to create pieces that would be impossible, or nearly so, using conventional design and sculpting techniques. His 20.9"×13.4"×13" piece, *Agrieborz* (right), looking like a medical manual illustration of an alien nervous system, took nearly six months to design and was hugely challenging to print. After that success, he decided to push himself further with *Nesurak* (40.9"×19.3"×21.3"), another figure. For this piece, he used many individual components (between 200–300) that he processed separately after printing. The resulting piece looks like a robot samurai bust that some futuristic overlord might proudly display on an office shelf. Ervinck spent nearly 2,000 hours on the components after printing them to achieve the retro-futuristic, biomech look that he was after. In Ervinck's most recent series, *Skin Mutations* (left), he continues to explore fluid forms, but the objects in this series contain more discernible component forms within the flowing shapes.

Ervinck's work has caught the attention of 3D printing companies Materialise and Stratasys. He has been collaborating with both firms in his never-ending desire to push the artistic boundaries of 3D fabrication tools into new and exciting territory.

—*Gareth Branwyn*

Nick Ervinck, Luc Dewaele

WOODEN WONDERS

TOMSKY.CO.UK/GALLERY

Martin Tomsky is "frequently distracted by events that didn't happen in places that don't exist." Luckily for us though, he has been able to turn these distractions into beautiful pieces of artwork to share with the world.

Tomsky considers himself more of an artist than a maker. He went to art school to study illustration and just happened to end up in a model making woodshop in 2012 cutting pieces for architectural models. He quickly realized that his art would translate magnificently into woodcut pieces. And from there his first woodcut artwork, *The Loneliness of Charon*, was born.

Since then Tomsky has made work of various scales and experimented with color and wood stains. But all of his larger pieces have one thing in common: they tell a story. "Most of the stories in my personal work are from my imagination, but often referring to or inspired by books that have left an impression on me," he says. His influences include Tolkien, Ursula K. Le Guin, China Miéville, H. P. Lovecraft, Iain M. Banks, Philip K. Dick, *Akira*, and *Spirited Away*.

Once he has an idea, extensive planning goes into each piece. He starts with a scribble in his notebook just to figure out where everything will be placed. He then makes another sketch at a larger size, closer to scale. He does another round of drawing, adding in final details, and then scans the drawing and uses illustrator to manually vectorize the image. From there he fine-tunes the design, layering, and coloring. Then he cuts the shapes out of high quality plywood, stains using custom staining mixtures, and assembles.

All said and done some of his larger projects have taken months to complete. Tomsky finds drawing up the vectors to be the most time consuming step and assembly and finishing to be the hardest. He says he can go through two entire audiobooks in the course of one piece.

When asked what's coming next Tomsky says "I have just finished designing a new jewelry collection inspired by classic fantasy and I have a longer personal project of creating my own world that will hopefully culminate in an exhibition." —*Sarah Vitak*

Martin Tomsky

DREMEL® DIGILAB

advanced technology made simple

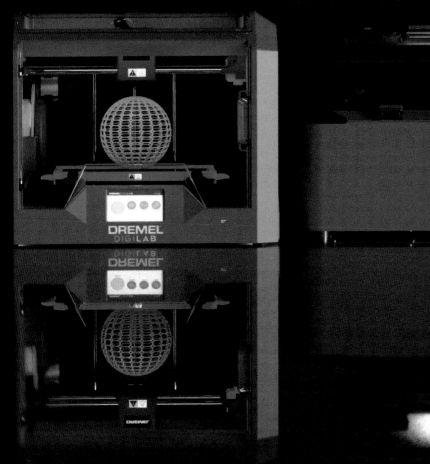

A SPATIAL PLACE

THEVERYMANY.COM/SUZHOU-BIENNALE

There is nothing like experiencing a piece of art that transports you to another place. And that is exactly what *Boolean Operator* does. **Marc Fornes / TheVeryMany** studio created the piece for the Suzhou Jinji Lake Biennial event, aiming to evoke a sense of being immersed in the magical *Extraordinary Voyages* of Jules Verne.

The team worked at a fast and furious pace, ultimately taking "ten days for strategy up to production files. Ten days for cutting and painting of the parts. And a week of assembly by up to 15 people the last few days." The project is a prototype to test production processes in anticipation of a larger and more permanent piece to come.

To create it, they first worked up a mesh of the overall shape — in this case using the intersection (or Boolean) of spheres. From there they computationally broke up the mesh into thin strips that could be cut on flat sheets. The final step in designing the structure was to run a "crawling" search protocol that takes into account the whole structure and adds the beautiful holes and cutout motifs throughout.

The work begins to exist in the real world as it enters the digital fabrication steps. The strips were cut out of thin (1mm–2 mm) aluminum with a CNC router. *Boolean Operator* is made of 1,673 individual strips. The team rivets the pieces together and the bends and folds naturally take form — without any manual folding.

The team says that even though they place a lot of focus on technology and design, "the piece lives in the world and is activated by its users — usually, very curious kids — then we know if it's truly a success. Children are the best investigators; they find all the climbable bits, the spaces to hide, to run around. And when their parents finally catch up to them, they too take notice of the structure." —*Sarah Vitak*

NAARO

Ben Semisch

BRITE DONE RITE JASONWEBB.IO

There are few toys from the '60s and '70s more fondly remembered than Hasbro's Lite-Brite. Who from that era doesn't recall the TV jingle? "Lite-Brite, making things with liiiiight. Out-of-sight, making things with Lite-Brite."

Minneapolis-based "creative technologist" **Jason Webb** managed to invoke much of "the magic of colored lights" from that iconic childhood toy with his *Giant Lite-Brite* installation, built in 2017 for the light exhibition at KANEKO, an art space in Omaha, NE.

Webb meticulously documented the process of creating his free-standing, lighted pegboard with four interactive surfaces in a detailed build log on his website. His installation was inspired by Noah Weinstein's jumbo Lite-Brite board project posted years ago on Instructables.

The multi-colored light pegs are made of cut acrylic rod and they feeling that they are plugging in light bulbs as they push the pegs into the holes. In reality, the peg "lights" up after being pushed beyond the black baffle and is exposed to the LED tube lights inside the structure. In total, 4,600 holes cover the four active surfaces of the piece and 4,600 colored pegs were made. The main structure for the display was constructed of welded-steel square tubing and CNC-cut birch plywood.

"Throughout the build process, I was able to apply familiar skills in CAD, CNC routing, laser cutting, and 3D printing, and to pick up some new ones, like MIG welding and metalworking," says Webb. "Any project where I get to learn something new is a win in my book!"

After the KANEKO exhibition, the piece was donated to the Autism Center of Nebraska. Kids will be making things with *Giant Lite-Brite*

MAKE THEIR HOLIDAYS
BRIGHT

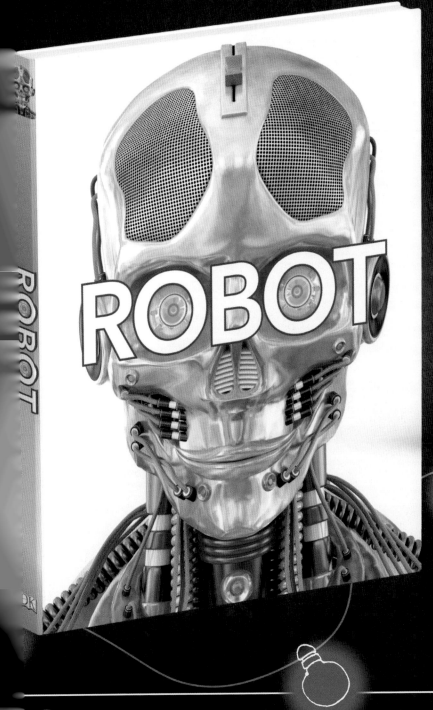

ROBOT

JUNIOR MAKER

Experiments to try, crafts to create, and LOTS to LEARN!

STAR WARS MAKER LAB

20 CRAFT AND SCIENCE PROJECTS

Build · Experiment · Create · Discover

Baby ROBOT

A beep-buzz, light-up story!

 A WORLD OF IDEAS:
SEE ALL THERE IS TO KNOW

www.dk.com

David Handschuh, Ben Hider, Andrew Kelly, Mike Sense

FRESH, FUN, AND FABULOUS

In its ninth year, **World Maker Faire New York** showcases new ideas and a sense of wonder for attendees

There's always a mix of new exhibits and returning favorites at the New York Hall of Science, and this year's event was no different. It's next to impossible to show the incredible variety of all the projects, presentations, and performances that take place over the course of two days, but here are a few of the highlights.

1. Artist Christian Ristow's *Hand of Man* was a crowd favorite. The enormous hydraulic hand is powerful enough to pick up full-sized cars. Lucky attendees got the chance to take control using a special glove that allowed the piece to mimic their movements.

2. We couldn't resist this cosplay photo op with the Long Island R2-D2 Builders Group and special guests.

3. Eepy Bird's Diet Coke and Mentos show is always a crowd pleaser, explaining the science behind making a huge, fun mess.

4. The Fat Cat Fab Lab from NYC's West Village brought a giant rotating music sequencer. Colored wooden pegs triggered different sounds and an enormous lever off to the side controlled the tempo.

5. Hailing from NCC Fab Lab in Bethlehem, PA, *Tobor* is a giant robot dinosaur that doubles as a giant robotic gripper. A sensor-laden gesture control glove sends commands wirelessly to the bot.

6. A number of musical acts roamed the grounds delighting attendees, but only this bari sax player was spouting fire in time to the music.

7. Artist Taewoo Park's *TV Being 009* is part of a series of pieces that uses broken and discarded electronic devices to meditate on the meaning of machine consciousness.

8. The souped up kiddie rides of the Power Racing Series may look comical, but these engineering powerhouses are no joke.

9. New this year, the DIY Content Creator Stage brought together some of the best makers on YouTube to talk about projects, making videos, and more. The audience was packed each day, and half the fun was watching the speakers, also often fans of their fellow panelists, get to interact with each other.

[+] See more editor favorites and video walkthroughs of the faire at makezine.com/go/wmfny18-live.

UNDER PRESSURE

A team of engineering grads turn a college waterjet cutter project into a startup venture

Written by Nisan Lerea

I OFTEN PLAYED WITH LEGO AS A KID GROWING UP IN RIVERDALE, N.Y., BUT MY FIRST MAKER PROJECT wasn't until high school when I built a 30-foot-tall trebuchet and competed in the 2006 Punkin Chunkin competition in Delaware. We hurled a 4-pound pumpkin 531 feet — good enough for third place in the Youth Trebuchet division. As I watched 800 pounds of counterweight launch the pumpkin through the air, I knew I wanted to study engineering.

RACING TO LEARN

I met my future Wazer co-founder Matt Nowicki during orientation of my freshman year of college at the University of Pennsylvania. All of the extracurricular clubs had lined the main walk on campus and were trying to recruit new members. Parked on the side of the path was an open-wheeled, Formula 1-style race car. Matt, a senior and clearly the leader of the club, explained that the Formula SAE Team builds a new car each year and races it in an intercollegiate competition in Michigan. I didn't love cars, but I knew this is where I would really learn how to make things. I immediately joined the team.

I would go on to spend hundreds of hours in Penn's machine shop, CNC-milling metal parts for the race car, for research labs, or for my own coursework. Because of all the setup and breakdown time that machining necessitates, I would regularly work late into the night making a part. Whenever possible the engineers would avoid the shop because of the time commitment and instead would design parts that could be laser-cut, which was way faster. The downside to using a laser was the parts had to be made in acrylic or MDF, because, as is the case with lasers at most makerspaces, ours could only cut certain soft materials.

SENIOR PROJECT

What we really needed was a waterjet for cutting sheet metal, but Penn never had one because they were so big and expensive. So in 2011 my professor suggested that we attempt to build a small waterjet for our yearlong senior design project. I loved the idea for many reasons: it was an engineering challenge, it involved my passion for making things, and I knew that there was real potential for the product — and I always wanted to be an entrepreneur. By May of 2012, our team had built the first small-scale waterjet, capable of cutting through ¼" aluminum and ⅛" steel.

Then we all graduated. I felt this could be more than a school project, but I figured it would be beneficial to get some real-world experience working as an engineer before embarking on this adventure. Fortunately for me, Matt, who had been working for the hardware startup BioLite in Brooklyn, N.Y., called me and said they were looking to hire a mechanical engineer. I signed up. I worked there for two years and was involved in two complete product-development cycles designing portable camping gear.

In 2014 Hackaday somehow got wind of our senior design waterjet project and published a blog post about us. Hundreds of people emailed us asking if we had plans to commercialize the technology. It was eye-opening for me, because it wasn't just engineers who were asking. Artisans, makers, and small businesses of all sorts inquired as well.

DIVING IN

By 2015 I was ready to make the leap to start a waterjet company. But I needed a partner. Luckily, Matt, who had since moved on from BioLite, was looking for a change. It wasn't hard to convince him to join me as co-founder and CTO.

We started out by researching the market from my parents' basement and testing the Penn waterjet prototype in the backyard. Then we googled "hardware accelerator" and discovered Hax, an accelerator for hardware startups in Shenzhen, the electronics capital of the world. We joined Hax in January 2016, hired Dan Meana and Christian Moore — two engineers from the Penn race car team — and moved to China.

We were amazed by the speed and affordability of prototyping at Hax. Wazer utilizes a lot of off-the-shelf hardware like hoses, fittings, valves, solenoids, and motors. We found that commodity hardware near Shenzhen was roughly one-tenth the cost of that in the U.S., often for the exact same parts sold by McMaster-Carr, Digi-Key, or Amazon. At Hax we were surrounded by entrepreneurs with whom we could share ideas and receive unbiased feedback.

21ST CENTURY BAND SAW

We launched Wazer on Kickstarter eight months later, in September 2016. The campaign was a huge success. We knew it would be difficult to transition from prototype to production. After qualifying vendors and redesigning the machine for volume production, our team established Wazer's headquarters in the Brooklyn Navy Yard, where we are part of a mission to revitalize manufacturing in New York City. Every Wazer is built in our combined office/workshop/assembly facility and is thoroughly tested prior to shipment. It took an additional 21 months — twice as long as expected — before we delivered the first Wazers to our extremely patient customers.

I certainly wish I had a waterjet back in high school and in college when we were building the trebuchet and our race car. Wazer is the 21st century band saw, a digital cutting tool that belongs in every workshop, because it cuts every material. Six years after its initial inception I am most excited to see the amazing and varied things that our customers make with Wazer — creations that no one on our team ever dreamed of. ◢

Mark Madeo, Josh Itzkowitz, OFPD (Office for Product Design), Courtesy Nisan Lerea, Adobe Stock - Vjom

① A peek under Wazer's cover shows its 12"×18" cutting area and nozzle.

② Pennies, with the metal around Lincoln's head cut out by Wazer, create a unique jewelry material. Design by Stacey Lee Webber.

③ Demonstrating quick, clean cuts in metal.

④ Lerea and his Penn teammates show off the first iteration of their desktop waterjet cutter.

NISAN LEREA, co-founder and CEO of Wazer, is a maker at heart and loves finding creative solutions to the problem at hand.

2019 DIGITAL FABRICATION GUIDE

Olansky Luciano

Chris Yohe

Ryan Priore

Kelly Egan

Matt Stultz

Matt Dauray

Jen Schachter

Mandy L. Stultz

TECH TOOLS

New digifab machines and a new category give you more project options than ever before *Written by Matt Stultz*

Kelly Egan

WELCOME TO THE 2019 *MAKE:* DIGITAL FABRICATION GUIDE! For the seventh straight year, our group of talented, dedicated, and kind of crazy digital fabrication experts gathered to put as many machines as possible through the paces to bring you a complete understanding of the latest maker tools. Assembling at the Ocean State Maker Mill (oceanstatemakermill.org) in Rhode Island, we used the historic, industrial factory space to thoroughly test the latest in 3D printers, CNC machines, and laser, vinyl, and — for the first time in our reviews — waterjet cutters. We'll help you find the right machine for your needs in the ensuing pages, with reviews, scores, data charts, and category selections. And you can find even more machine comparisons in our companion online guide: makezine.com/go/3dp-comparison. Enjoy. ●

MEET THE TEAM

Everyone on this year's testing team is a veteran of past *Make:* digifab shootouts, and all are members of makerspaces where we use these tools and teach them to others. We love what we do!

Kelly Egan is an artist and maker living in Providence, RI and a member at Ocean State Maker Mill.

Olansky Luciano is a hobbyist, a maker, 3D printer and CNC machine enthusiast. A child at heart who is currently finishing his mechanical engineering degree, he is a member at Ocean State Maker Mill.

Jen Schachter is an ex art-schooler, community project builder, maker movement researcher, safety manual author, tiny house dreamer, best ideas in the shower-er. Nation of Makers, Tested.com, and We the Builders collaborator. RWD Foundation Fellow, Open Works Baltimore resident.

Chris Yohe is a software developer by day, digital fabrication maestro by moonlight. Dad. Rugby player. Espresso lover. Co-founder of 3DPPGH. HackPGH member. Current machine count redacted.

Mandy L. Stultz pays the bills as an online-marketer but her love is making. Woolly crafts and figuring out how many different ways she can make her own T-shirts are priority. Two dogs and a cat-flavored familiar. She likes hoods, glitter vinyl, and fake fur.

Matt Stultz is *Make:*'s digital fabrication editor in charge of heading up this team and is the founder of 3DPPVD, Ocean State Maker Mill, and HackPGH.

Matt Dauray is a mechanical engineer and materials connoisseur. Finish carpenter/leatherworker, Matt spends his spare time at the Ocean State Maker Mill producing dust and design for furniture, leather jewelry, motorcycle parts, and other tech-based manufacturing.

Ryan Priore is a spectroscopist and entrepreneur in the photonics industry. He is the Assistant Cubmaster for Pack 344, an active member of F3 Pittsburgh, a member of HackPGH, and the co-founder of 3DPPGH.

WATERJET CUTTERS

Introduce exceptional versatility to your workshop with this exciting new type of desktop machine

Written by Matt Stultz · Illustrated by Rob Nance

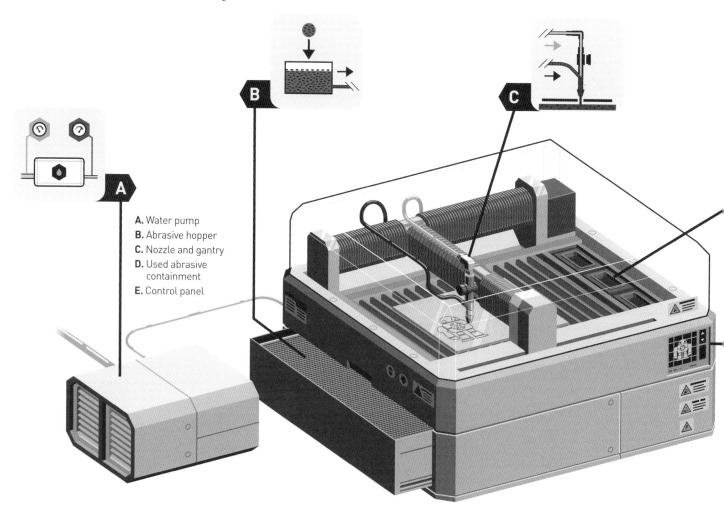

A. Water pump
B. Abrasive hopper
C. Nozzle and gantry
D. Used abrasive containment
E. Control panel

WATERJET CUTTERS, LIKE 3D PRINTERS IN THE PAST, HAVE LONG BEEN RELEGATED TO ONLY THE HIGH-END, professional shops that could afford them. That's all changing with a new batch of waterjets that are targeting makers and their organizations, bringing the cost of ownership to an attainable level.

HOW THEY WORK

A waterjet is a subtractive tool used to cut through almost anything. It is especially handy for materials that are challenging for other tools, like hard metals, carbon fiber, and glass. A waterjet uses extremely high pressure and high velocity water combined with a fine powder abrasive (usually garnet) to quickly etch through your material to produce clean, even separation.

The basic mechanics of a waterjet

are very similar to those of other digital fabrication machines. Stepper motors move a gantry on an x- and y-axis, shifting the cutting head around the material. A motor driven z-axis isn't needed because the nozzle height can be manually set at the beginning of the operation. The workflow is very similar to that of using a laser cutter.

To prevent the waterjet from cutting through itself, the workpiece sits above a tank of water that dissipates the power coming from the stream of water. There is also some kind of sacrificial surface for the material to sit on and be clamped to.

SOME ISSUES

While having the power to cut through any material feels amazing and opens up totally new project possibilities, there are some disconcerting aspects to using a waterjet. High pressure water spraying around, even inside an enclosure, feels like a problem waiting to happen — especially when high voltage is in play for high pressure compressors.

The abrasive sand is also an issue. Sadly, this material is not reusable and its cost can add up quickly as you are making big cuts. The higher pressure the system, the less abrasive it needs to use to cut a part. And it's hard to use one of these machines without finding some of the abrasive in your mouth and all over your clothes. (Wear your safety glass — this shouldn't be considered optional around one of these machines, even in an enclosure).

Our testing focused on two big names in waterjet cutting: the ProtoMax, produced by Omax, who have been in the waterjet world for over 25 years, and Wazer, the new startup who gained attention with their crowdfunding campaign and relative low cost (read more about the company on page 16.) We hope to add to this list, with rumors of another machine possibly coming to market soon.

Have you used a waterjet on any of your projects? Share pictures of your creations with us at editor@makezine.com. We are excited to see what makers are doing and will do with this technology. ◐

Get Your Feet Wet

Want to get started with your waterjet cutter? Here are some project ideas to inspire you:

Steel Gaming Dice
Matthew Dockrey (twitter.com/attoparsec) provides a guide on creating oversized dice that are as cool as they are unique: instructables.com/id/Gaming-dice-made-from-waterjet-cut-steel-sheet-fol

DIY Carabiner
One of the more versatile accessories, maker Christian Reed explains how he made his unique magnet-closure 'biner entirely from scratch: instructables.com/id/Magnabiner-Build-your-own-Carabiner

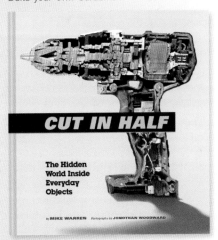

Cut in Half
Waterjet master Mike Warren bisects items on his YouTube channel (youtube.com/c/cutinhalf) to viewers' delight. He now documents his work in his new book *Cut In Half: The Hidden World Inside Everyday Objects*, peeling open everything from golf clubs to digital cameras with the precision that only a waterjet cutter can provide. chroniclebooks.com/titles/cut-in-half.html

WAZER

This reasonably priced benchtop machine really delivers
Written by Matt Stultz

WHY TO BUY
This is the first easy to use, relatively affordable waterjet. If you need parts made from difficult-to-cut materials, it is really hard to go wrong with the Wazer.

PRO TIPS
Pay close attention to how much spray you get when cutting the first demo part. In the future if you are getting a lot more spray it could be because you haven't pierced the material correctly and you need to correct your settings.
    Make your life easier: Order the leg package so you don't need a sturdy table to put it on.
    The plastic operating oil plug on the pump is very easy to accidentally cross thread so be careful and take your time when swapping with the travel plug.

■ **WEBSITE**
wazer.com
■ **MANUFACTURER**
Wazer
■ **BASE PRICE**
$4,499
■ **ACCESSORIES INCLUDED AT BASE PRICE**
Abrasive buckets (2), Water sensors (2), plumbing kit, spare cutting bed
■ **MAXIMUM CUTTING DIMENSIONS**
305×457mm
■ **WORK UNTETHERED?**
Yes, SD card
■ **ONBOARD CONTROLS?**
Yes, control pad with LCD
■ **MAX PRESSURE**
Undisclosed
■ **JOB SETUP SOFTWARE**
WAM (Wazer CAM)
■ **OS**
Web-based so fully cross platform
■ **OPEN SOFTWARE?**
No
■ **OPEN HARDWARE?**
No

WE HAD BEEN ITCHING TO GET OUR HANDS ON A WAZER SINCE ITS Kickstarter campaign launched, and have not been disappointed.

Setting up the machine is fast and easy using quick-connect hoses, and it comes with parts to either permanently plumb it or hook it up to your sink. Heavy to begin with, once it's full of water and abrasive it will be around 300lbs, so help and planning is key: Make sure it is close enough to your water supplies and drainage.

SIMPLE, UNTETHERED CUTTING

Wazer has made an easy-to-use web app CAM, called WAM (a name we can't even type without thoughts of '80s music videos). It could be a little more full-featured — color mapping for different processes in your file would be huge — but is highly intuitive. To set up your job, simply import a SVG or DCF file, scale and rotate the file if needed, select the cutting material, choose the cutting path (outside, inside, or centerline), decide if you want tabs or lead ins, then export your G-code. Caution: Parts layout is absolute, not relative — many CNC machines let you set your own home position, the Wazer will not.

A built-in SD card slot and LCD screen with control buttons means there's no need to leave a computer hooked up, a feature we really appreciate in a machine using water. All your interactions can be run from the onboard menu, and once it gets going, you just need to keep the hopper filled with abrasive. One way that Wazer has brought the cost down so drastically is by using a lower pressure pump, which means it needs more abrasive and the hopper needs to be filled regularly (and the waste vats emptied). Note: the Wazer stops every hour for abrasive fill-up, whether it needs it or not.

ACCESSIBLE AND AFFORDABLE

As you would expect from a waterjet, the cuts were clean and ready to be used straight out of the machine. While the operational cost of bigger waterjets is much lower as abrasive isn't cheap, we think that for low-volume usage, you can't beat the ease and price of the Wazer. ◉

AS YOU WOULD EXPECT FROM A WATERJET, THE CUTS WERE CLEAN AND READY TO BE USED STRAIGHT OUT OF THE MACHINE

PROTOMAX

Compact and capable, perfect for universities or
makerspaces *Written by Matt Stultz*

- **WEBSITE**
  protomax.com
- **MANUFACTURER**
  Omax
- **BASE PRICE**
  $19,950
- **ACCESSORIES INCLUDED AT BASE PRICE**
  Maintenance tools, laptop with software installed,
  55lbs of abrasive
- **MAXIMUM CUTTING DIMENSIONS**
  305×305mm
- **WORK UNTETHERED?**
  No
- **ONBOARD CONTROLS?**
  No, power switch only
- **MAX PRESSURE**
  30,000psi
- **JOB SETUP SOFTWARE**
  Proto Layout, Proto Make
- **OS**
  Windows
- **OPEN SOFTWARE?**
  No
- **OPEN HARDWARE?**
  No

WATERJETS HAVE BEEN BIG AND EXPENSIVE BUT INDUSTRY LEADER

Omax is trying to reach a new market with
a smaller machine and a smaller price
point (relatively). The ProtoMax is a more
accessible waterjet for customers who don't
need to cut large objects but still want a
production-ready machine.

SMALL AND STURDY

At just under $20K, probably most of us are
not going to have a ProtoMax in our garage,
but for makerspaces and universities
with a legit budget it gives some serious
functionality for your members/students.
Also this is an all-in price — it comes with
everything, even a laptop with the software
pre-installed.

This machine is heavy and cumbersome;
even though it's on wheels, have a couple of
helpers on hand when it arrives. The video
installation guide annoyingly required lots
of stops and rewinds, but once it was all
together the machine felt not only sturdy
but professional.

The full-featured Inteli-Max Proto Layout
even includes drawing tools if you want to
create a quick part and skip a CAD package
altogether. It could be a little more intuitive
though: submenus are hidden under icons
and accessible through right-clicks, and the
part-making process is not listed clearly.
After the layout is complete, Inteli-Max
Proto Make sends that job to the machine.

When you start up the machine for the
first time, you will need to go through a
couple test and warm-up procedures. This
will also partially fill up the machine's catch
tank; you will need a hose (the good way) or
a bucket (the tiresome way) to fill the rest of
the tank when the tests are complete.

HIGH-QUALITY CUTS

Your work is held down with a set of provided
spring clamps, which worked really well and
were easy to operate. Cuts on the ProtoMax
were crisp and great. I tested on ⅛" steel
and aluminum and had no problems with
either material. The pressure coming out of
the ProtoMax's pump is evident in how the
parts came out. This machine packs a big
punch in a small footprint. ⊘

THE PROTOMAX IS FOR CUSTOMERS WHO DON'T NEED TO CUT LARGE OBJECTS BUT STILL WANT A PRODUCTION-READY MACHINE

Kelly Egan

3D PRINTERS

This maturing market continues to innovate and evolve

Written by Matt Stultz · Illustrated by Rob Nance

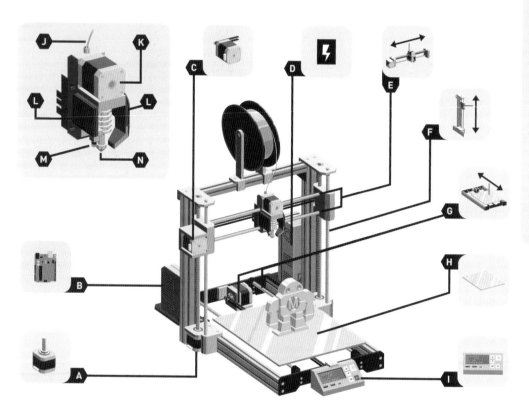

Ready Set Extrude

Getting started with your 3D printer? Try these hot projects:

Fortnite Boogie Bomb
Dead-man switch, sound board, RGB LEDs — Andrew Lamson's prop looks and sounds like the in-game grenade. Dance til you die! thingiverse.com/thing:2801202

Chain Mail Fabric
3D guru Agustin Flowalistik's new trick: printable fabric in rectilinear links. Coming to a cosplay near you. thingiverse.com/thing:3096598

A. Z-axis stepper motor
B. Motherboard
C. X-axis stepper motor
D. Power source
E. X-axis belt and shaft
F. Z-axis lead-screw and shaft
G. Y-axis stepper motor, belt, and shafts
H. Print bed
I. Control panel
J. Filament
K. Extruder stepper motor
L. Cooling fans
M. Extruder
N. Nozzle

PRINTER TESTING

We made only a minor change to our testing models this year, thickening the Full Bed Dimensional Accuracy probe to help it keep its shape better. The rest of our probes and processes have remained the same, with digital and physical redundancies to ensure nothing is missed. To learn more about how we test, visit makezine.com/comparison/3dprinters/how-we-test/shootout

2018 HAS BEEN A BIT OF A SHAKE-UP YEAR FOR THE 3D PRINTING INDUSTRY, seeing big players like Printrbot close their doors and a few others tightening their belts. Some tech observers have seen this and taken it as a sign that the 3D printing industry is failing and it's time to move to the next big thing. A careful observer will see that sales are still going strong, new users are entering the market every day, and those companies who have found their place are digging in for the long run. The startup hype may be over but now it's time for a maturing industry to get the real work done.

This isn't to say that there are not exciting things happening. Machines are getting better, more cost effective, and more configurable, and more accessories are coming to the market. MatterHackers offers a Dell Computers-like buying experience for their Pulse printer, allowing you to select the features you want. LulzBot has taken advantage of their recent switch to E3D hotends by releasing an optional fine-detail tool head months after the LulzBot Mini 2 launched. Prusa's new MMU2 adds a ton of functionality to their already feature-rich machine. This, of course, is just a short list but enough to get enthused about not only where we are, but where things are going. ⊘

PRUSA I3 MK3

Slick upgrades further elevate this top-notch machine

Written by Olansky Luciano

PRUSA RESEARCH EXCELS AT TWO THINGS: CREATING ROCK STAR PRINTERS AND REFINING THEM based on feedback. Last year we got a peek at an early MK3, but now that we've had our hands on a full production unit for this year's tests we can confidently say the Prusa team knocked it out of the park.

KEY UPDATES

The extruder has been upgraded to Bondtech drive gears; it can reliably extrude exotic and flexible filaments. The filament detector will pause your job to fix any feed issues — we cut the filament during a test print and recovered without skips or shifts. The removable PEI bed not only makes it easy to get prints to stick but even easier to remove when the print is complete. The option of purchasing additional beds means turnaround times when doing production work with the MK3 will be fast and easy.

The MK3 has full power-loss recovery built-in, continuing where it left off when interrupted. While testing one of the waterjets, power in our space went out; when it came back the printer rebooted and asked if we'd like to resume, then picked right back up. Magic.

DYNAMIC TRINAMICS

The Einsy Rambo motherboard with Trinamic drivers has made the MK3 so reliable that users of any skill level will have a great experience. The drivers make the printer faster as well as quieter — coupled with its Noctua fan, we could hardly hear it running in the room. An educator running an MK3 in a classroom won't find the students distracted by the noise, and you won't need to force your neighbors to listen to your Netflix binge just so you can hear the TV over your printer when it's in your living room. If anything bumps into the print head while operating, the drivers will detect the resistance, triggering a home and restart, nearly eliminating layer shift.

It's hard to believe that this feature-rich machine still comes in at under $1,000. The Prusa team just keeps raising the bar. ⊘

2019 DIGITAL FABRICATION GUIDE
Make:
Editor's Choice
★ ★ ★
Prusa MK3

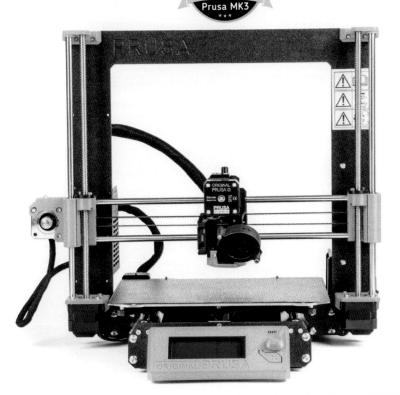

MACHINE RATING

	1	2	3	4	5
VERTICAL SURFACE FINISH	■	■	■	■	
HORIZONTAL FINISH					
DIMENSIONAL ACCURACY					
OVERHANGS					
BRIDGING	■	■	■		
NEGATIVE SPACE					
RETRACTION					
SUPPORT MATERIAL					
SQUARENESS	■	■	■	■	
FULL BED ACCURACY					
Z WOBBLE	PASS				

Price as Tested **$999**

46

■ **WEBSITE**
prusa3d.com

■ **MANUFACTURER**
Prusa Research

■ **BUILD VOLUME**
250×210×210mm

■ **BED STYLE**
Heated/MK52 removable magnetic spring steel sheet with PEI surface

■ **FILAMENT SIZE** 1.75mm

■ **OPEN FILAMENT?** Yes

■ **TEMPERATURE CONTROL?**
Yes, extruder (300ºC max); bed (120ºC max)

■ **PRINT UNTETHERED?**
Yes, SD card

■ **ONBOARD CONTROLS?**
Yes, integrated LCD and SD card controller

■ **HOST/SLICER SOFTWARE** Slic3r Prusa Edition; PrusaControl

■ **OS** Mac, Windows, Linux

■ **FIRMWARE** Open source Prusa Marlin

■ **OPEN SOFTWARE?** Yes, Slic3r PE is GNU AGPL-3.0, PrusaControl is GNU GPL-3.0

■ **OPEN HARDWARE?**
Yes, GPL

PRO TIPS

Order an extra PEI coated bed for quick and easy changes between prints. It will save you time and frustration if your print is stuck too well.

WHY TO BUY

The MK3 is the best fully loaded 3D printer in its price range, and offers 24/7 customer support, extended community, and easy to use software for beginners or experts.

PRINT

Kelly Egan

PRUSA I3 MK3 WITH MMU2

Increase your filament capabilities with this updated multimaterial addition *Written by Ryan J. Priore*

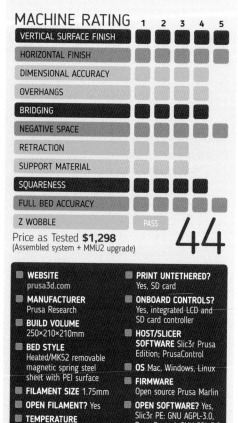

MACHINE RATING

	1	2	3	4	5
VERTICAL SURFACE FINISH					
HORIZONTAL FINISH					
DIMENSIONAL ACCURACY					
OVERHANGS					
BRIDGING					
NEGATIVE SPACE					
RETRACTION					
SUPPORT MATERIAL					
SQUARENESS					
FULL BED ACCURACY					
Z WOBBLE	PASS				

Price as Tested **$1,298**
(Assembled system + MMU2 upgrade)

44

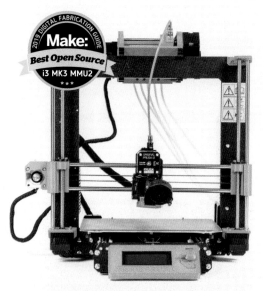

2019 DIGITAL FABRICATION GUIDE
Make:
Best Open Source
i3 MK3 MMU2
★★★

WHY TO BUY

This versatile and uncompromising addition allows you to print both flexible and non-flexible filaments as well as the original MK3 with less waste. Switching from single to multiple filaments is a trivial process from the PrusaControl slicer.

THE MULTI-MATERIAL UPGRADE 2 (MMU2) IMPROVES THE ORIGINAL MMU
and adds an extra (5th) material option. Ours arrived pre-installed on an i3 MK3, although you can't currently buy the two assembled — it's just an upgrade option for now.

The most obvious improvement is the single MK3 direct drive extruder connected to the head unit via a single PTFE tube, offering a constrained path for filament loading and unloading from the unit to the extruder. This translates to better printing with both flexible materials and traditional filaments. The head unit contains three stepper motors for controlling the intricate loading/unloading process. It also includes filament sensors in the head unit and the extruder.

In testing, the MMU2 had almost no effect on the stock MK3's performance; the only real difference is the prompt to select a filament line. One could keep up to five materials loaded for on-demand use, or even operate in redundancy mode. ⊘

WEBSITE prusa3d.com

MANUFACTURER Prusa Research

BUILD VOLUME 250×210×210mm

BED STYLE Heated/MK52 removable magnetic spring steel sheet with PEI surface

FILAMENT SIZE 1.75mm

OPEN FILAMENT? Yes

TEMPERATURE CONTROL? Yes, extruder (300°C max); bed (120°C max)

PRINT UNTETHERED? Yes, SD card

ONBOARD CONTROLS? Yes, integrated LCD and SD card controller

HOST/SLICER SOFTWARE Slic3r Prusa Edition; PrusaControl

OS Mac, Windows, Linux

FIRMWARE Open source Prusa Marlin

OPEN SOFTWARE? Yes, Slic3r PE: GNU AGPL-3.0, PrusaControl: GNU GPL-3.0

OPEN HARDWARE? Yes, GPL

N2 PRO2

A major redesign delivers a leap forward in quality and comfort
Written by Chris Yohe

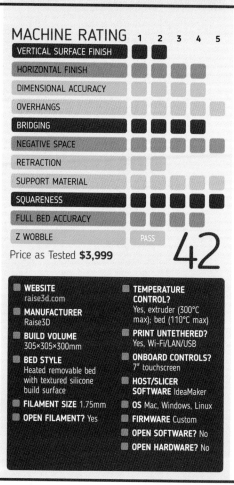

MACHINE RATING

	1	2	3	4	5
VERTICAL SURFACE FINISH					
HORIZONTAL FINISH					
DIMENSIONAL ACCURACY					
OVERHANGS					
BRIDGING					
NEGATIVE SPACE					
RETRACTION					
SUPPORT MATERIAL					
SQUARENESS					
FULL BED ACCURACY					
Z WOBBLE	PASS				

Price as Tested **$3,999**

42

2019 DIGITAL FABRICATION GUIDE
Make:
Best Large Format
N2 Pro2
★★★

WHY TO BUY

Raise3D has released a professional quality printer that builds their previous success into an even better machine, the Pro2, with tons of upgrades under the hood.

RAISE3D FOLLOWED UP A TEAM FAVORITE WITH A DECEPTIVELY SIMILAR
offering that is bursting with new features, while still packing their award-winning quality. The Pro2 shows just how seriously they take feedback, building in now-standard community mods as well as totally new enhancements.

In our tests, the ideaMaker software was easy to use, and Wi-Fi printing was a dream. We used Raise3D Premium PLA and found filament loading and unloading to be a breeze. Among the many changes you will find a new 32-bit control board, camera, dual Bondtech-style extruders, HEPA filter, redesigned magnetic build platform, and even a new endstop design.

With upgrades to please everyone from the extruder savvy to the fume conscious, the Pro2 is as close to set-it-and-forget-it as we've seen in a top level machine. For those looking for a printer that bridges the maker and maker pro gap — here it is! ⊘

WEBSITE raise3d.com

MANUFACTURER Raise3D

BUILD VOLUME 305×305×300mm

BED STYLE Heated removable bed with textured silicone build surface

FILAMENT SIZE 1.75mm

OPEN FILAMENT? Yes

TEMPERATURE CONTROL? Yes, extruder (300°C max); bed (110°C max)

PRINT UNTETHERED? Yes, Wi-Fi/LAN/USB

ONBOARD CONTROLS? 7" touchscreen

HOST/SLICER SOFTWARE IdeaMaker

OS Mac, Windows, Linux

FIRMWARE Custom

OPEN SOFTWARE? No

OPEN HARDWARE? No

Kelly Egan

JELLYBOX 2

Build your 3D printing knowledge by putting together this quality kit *Written by Kelly Egan*

NEON COLORS, CLEAR ACRYLIC, AND TONS OF ZIP TIES MAKE THE JELLYBOX 2 stand out from other 3D printers, but like its predecessor, where it really shines is solid performance and as a great introduction to the 3D printing process.

TEACHING TOOL

The JellyBox 2 is sold as a kit and is intended as an educational experience. While ours came pre-assembled for this review, online documentation is thorough, including images, videos, and explanations.

A pre-flight test built into the firmware walks you through a number of key checks, an especially important feature with kit printers, and makes it much easier for young builders and inexperienced users to quickly troubleshoot any problems.

There are a few issues with the onboard controls, like having to scroll all the way to the bottom to get to the print from SD menu option. Also an unusual feature is that after preheating the printer will beep once and wait until you press the knob again before printing. If you are used to other printers that just start to print, you might be disappointed if you come back expecting it finished and find it waiting for your input to even start.

CLEARLY CAPABLE

The JellyBox 2's hardware, while unusual, is well-designed and sound. The printer is more rigid than some printers made out of extruded aluminum. It doesn't shake or rattle when printing and our test prints came out well. This version adds a second fan to the extruder which should help prevent print droop, although we did see a little droop on the greater angles of the overhang test.

There are also nice details like the double-hinged door with magnet stops that easily tucks away to one side. Imade3D also makes great use of the acrylic surface of the enclosure by including diagrams of the printers electronics and small rulers for running calibrations like extruder steps.

Overall the JellyBox 2's attention to detail and focus on team builds make this a great tool for learning the ins and outs of FFF printing.

MACHINE RATING	1	2	3	4	5
VERTICAL SURFACE FINISH	■	■	■	■	■
HORIZONTAL FINISH	■	■			
DIMENSIONAL ACCURACY	■				
OVERHANGS	■				
BRIDGING	■	■	■		
NEGATIVE SPACE	■				
RETRACTION	■				
SUPPORT MATERIAL	■				
SQUARENESS	■	■			
FULL BED ACCURACY	■		■		
Z WOBBLE		PASS			

Price as Tested **$949**

42

- **WEBSITE** imade3d.com
- **MANUFACTURER** Imade3D
- **BUILD VOLUME** 170×160×145mm
- **BED STYLE** Unheated. Upgrade to heated, multiple print surface options
- **FILAMENT SIZE** 1.75mm
- **OPEN FILAMENT?** Yes
- **TEMPERATURE CONTROL?** Yes, extruder (245°C max); upgraded bed (120°C max)
- **PRINT UNTETHERED?** Yes, SD card
- **ONBOARD CONTROLS?** Yes, integrated LCD and SD card controller with scroll wheel
- **HOST/SLICER SOFTWARE** Cura Imade3D Edition
- **OS** Mac, Windows, Linux
- **FIRMWARE** Marlin
- **OPEN SOFTWARE?** Yes, firmware is Marlin GPL, Cura is GNU Affero General Public License v3.0
- **OPEN HARDWARE?** Yes, CC BY-NC-SA 4.0

PRO TIPS
Be sure to wait for the OK dialog after the printer heats to confirm your print or you might be waiting forever for it to start printing.

WHY TO BUY
Learning how a 3D printer works can be half the fun of using it. JellyBox makes it easy to assemble your first 3D printer.

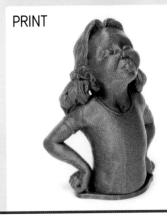

PRINT

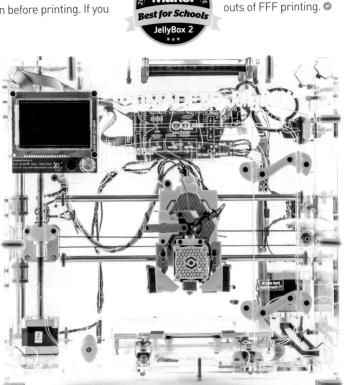

2019 DIGITAL FABRICATION GUIDE
Make:
Best for Schools
JellyBox 2
★★★

Kelly Egan

PULSE XE

High-end components and open source roots produce terrific results *Written by Ryan Priore*

MATTERHACKERS HAS BECOME THE ONE-STOP SHOP FOR 3D PRINTING hardware, accessories, and supplies. Their debut printer, the Pulse XE, is a who's who of best-in-class subcomponents culminating in an impressive open source printing experience.

SURPRISING STRENGTHS

Every other vendor sent us PLA filament to test with, but MatterHackers included what we considered a challenging filament, the company's own NylonX — consisting of 20% carbon fiber by weight. And it handled it like a champ; the surface finish of the test probes was exquisite. Make no mistake, the Pulse XE is more than capable of printing all of your favorite materials — it just makes printing nylon look so easy! The included PrintDry system also actively dries your hygroscopic (or water loving) filaments both prior to and during the printing process.

PRIMO PARTS

The Pulse line is based on the Prusa i3 design, with its open metal frame and 3030 extrusion. The brains of the Pulse XE are still an 8-bit Mini Rambo electronics board, while a sleek Panucatt Viki 2 LCD screen and BuildTak FlexPlate system are welcome upgrades. Vibration dampeners adorn the x-, y-, and z-axes for minimizing the transfer of motor-induced artifacts into the final prints, and a Bondtech extruder plus filament runout sensor provides a belts and suspenders approach to filament extrusion. The compact E3D hotend comes equipped with an Olsson Ruby nozzle for handling the abrasive filaments and a BLTouch for auto-tramming any build surface.

My one complaint is that I found the MatterControl software interface to be overwhelming and not intuitive, with attempts to blend basic design elements and traditional slicing capabilities. MatterControl is also required for setting the BLTouch probe offset, which is annoying if you need to adjust the offset in real-time or prefer to print from a microSD card.

Ultimately, the machine was a joy to use, and those looking to print functional prototypes with corresponding strength and durability will find the Pulse XE, much like MatterHackers, to be your one-stop shop. ⊘

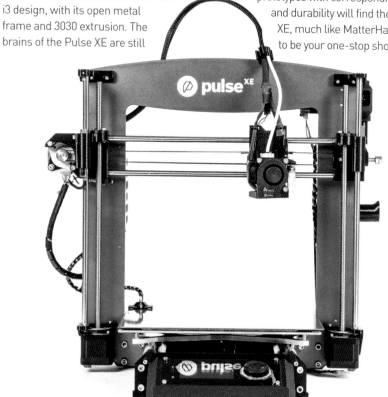

MACHINE RATING

	1	2	3	4	5
VERTICAL SURFACE FINISH					
HORIZONTAL FINISH					
DIMENSIONAL ACCURACY					
OVERHANGS					
BRIDGING					
NEGATIVE SPACE					
RETRACTION					
SUPPORT MATERIAL					
SQUARENESS					
FULL BED ACCURACY					
Z WOBBLE	PASS				

Price as Tested **$1,595**

42

WEBSITE
matterhackers.com

MANUFACTURER
MatterHackers

BUILD VOLUME
250×220×215mm

BED STYLE
Heated BuildTak FlexPlate with Garolite surface, removable magnetic spring steel sheet

FILAMENT SIZE
1.75mm

OPEN FILAMENT?
Yes

TEMPERATURE CONTROL?
Yes, extruder (300°C max); bed (120°C max)

PRINT UNTETHERED?
Yes, microSD, Wi-Fi

ONBOARD CONTROLS?
Yes, integrated LCD and SD card controller

HOST/SLICER SOFTWARE
MatterControl 2.0 Beta

OS Windows

FIRMWARE
Marlin github.com/ MatterHackers/ PulseV1Firmware/tree/ pulse_v2

OPEN SOFTWARE?
Yes github.com/ MatterHackers/ MatterControl

OPEN HARDWARE?
Yes github.com/ MatterHackers/ PulseOpenSource

PRO TIPS

Use the PrintDry system prior to printing if your filament has been exposed to the environment – and allow for space around it as it gets warm.

WHY TO BUY

The Pulse XE has a tremendous range of filament options. NylonX prints with ease and is the perfect complement to this open source beast.

PRINT

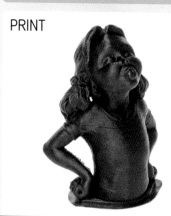

LULZBOT MINI 2
Refinements make an excellent machine even better
Written by Kelly Egan

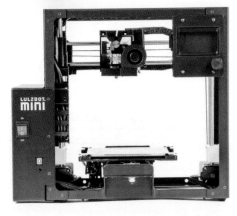

WHY TO BUY
The LulzBot Mini 2 is a well-built machine that makes a number of improvements on the original.

THE LULZBOT MINI 2 DOES A GOOD JOB MAINTAINING THE QUALITY of the original Mini while adding useful features.

The onboard control and SD card input are two of the more noticeable additions to the Mini 2, freeing up your computer from the potentially hours-long print process. It also now includes an E3D hotend, and the large hinged idler release found in LulzBot's other extruder has been replaced with a small tension knob and release lever, requiring a little less hand strength for loading and unloading filament. One feature that distinguishes the Mini 2 from many other printers including the original is the belt-driven z-axis. It did not noticeably decrease print time in our tests, although it should speed up operations like homing.

LulzBot's printers have a reputation for reliability and our initial tests did not contradict this. The Mini 2 would be a top choice for anyone looking for a machine they can count on. ◉

MACHINE RATING

	1	2	3	4	5
VERTICAL SURFACE FINISH					
HORIZONTAL FINISH					
DIMENSIONAL ACCURACY					
OVERHANGS					
BRIDGING					
NEGATIVE SPACE					
RETRACTION					
SUPPORT MATERIAL					
SQUARENESS					
FULL BED ACCURACY					
Z WOBBLE	PASS				

Price as Tested **$1,500**

42

- **WEBSITE** lulzbot.com
- **MANUFACTURER** LulzBot
- **BUILD VOLUME** 160×160×180mm
- **BED STYLE** Heated borosilicate glass/PEI
- **FILAMENT SIZE** 2.85mm
- **OPEN FILAMENT?** Yes
- **TEMPERATURE CONTROL?** Yes, extruder (290°C max); bed (120°C max)
- **PRINT UNTETHERED?** Yes, SD card
- **ONBOARD CONTROLS?** Yes, integrated LCD and SD card controller
- **HOST/SLICER SOFTWARE** Cura LulzBot Edition
- **OS** Mac, Windows, Linux
- **FIRMWARE** Marlin
- **OPEN SOFTWARE?** Yes, GNU Lesser General Public License Version 3
- **OPEN HARDWARE?** Yes, Creative Commons Attribution–ShareAlike 4.0 International Public License

---

MONOPRICE DELTA PRO
This sleek, sturdy full-size delta lives up to its name
Written by Chris Yohe

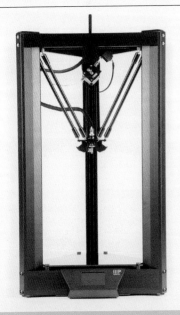

WHY TO BUY
Those looking to move from the mini to the full size delta experience, the Monoprice Delta Pro provides an easy to use, high-quality machine with great aesthetics.

THE MONOPRICE DELTA PRO IS A BIG STEP UP FROM ITS SMALL, INEXPENSIVE brethren, the MP Mini Delta, and both the user experience and design really fit the Pro name.

The sleek machine features linear rails and magnetic ball ends hold the hotend platform in place, which take advantage of Atom's MagSwap system, giving you access in the future to a whole range of tool heads, or even let you roll your own with their Module Development Kit. A 32-bit Lerdge-X control board provides great printing characteristics as well as an easy to use screen interface.

The test prints came out great, really showing off the Delta Pro's capabilities. We were also impressed with the speed at which this machine flew through the prints. If you are looking for a delta experience that doesn't feel miniaturized, check this one out — you may be pleasantly surprised. ◉

MACHINE RATING

	1	2	3	4	5
VERTICAL SURFACE FINISH					
HORIZONTAL FINISH					
DIMENSIONAL ACCURACY					
OVERHANGS					
BRIDGING					
NEGATIVE SPACE					
RETRACTION					
SUPPORT MATERIAL					
SQUARENESS					
FULL BED ACCURACY					
Z WOBBLE	PASS				

Price as Tested **$1,499**

41

- **WEBSITE** monoprice.com
- **MANUFACTURER** Monoprice
- **BUILD VOLUME** 270mm dia.×300mm tall
- **BED STYLE** Heated glass
- **FILAMENT SIZE** 1.75mm
- **OPEN FILAMENT?** Yes
- **TEMPERATURE CONTROL?** Yes, extruder (310°C max); bed (100°C max)
- **PRINT UNTETHERED?** Yes, USB drive, Wi-Fi
- **ONBOARD CONTROLS?** Yes, touchscreen
- **HOST/SLICER SOFTWARE** KissSlicer
- **OS** Mac, Windows, Linux
- **FIRMWARE** Custom
- **OPEN SOFTWARE?** No
- **OPEN HARDWARE?** No

Kelly Egan

ROSTOCK MAX V3.2

Major upgrades and new features solidify its crowd-favorite status

Written by Chris Yohe

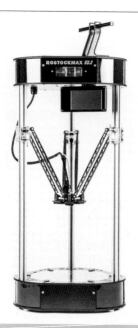

WHY TO BUY
SeeMeCNC has taken an old standard in their Rostock Max and have packed in some huge value adds in a new controller and hotend to make the Rostock Max v3.2 a highly recommended machine.

SEEMECNC IS BACK WITH A FRESH UPDATE ON A FAVORITE CLASSIC WITH the Rostock Max v3.2.

The upgrade to the SE300 assembly sees another generation of plug and play hotends coming to fruition, and auto leveling is now being driven by flexible strain gauges instead of accelerometer probes. The switch to the Duet Wifi 32-bit controller board, sporting much quieter Trinamics drivers and an onboard Wi-Fi module to enable easy printing from your PC, tablet, or phone, is by far the most important change. When powered by the great 5" touchscreen display (which also enables printing from an SD card) it becomes the tipping point that should get you out of your seat and in line to buy.

These and other new additions bring a huge dose of value to an already proven machine, and the print quality remains worthy of praise. We highly recommend the Rostock Max v3.2 to anyone looking for a delta printer. ◎

MACHINE RATING

	1	2	3	4	5
VERTICAL SURFACE FINISH	■	■			
HORIZONTAL FINISH	■	■	■	■	
DIMENSIONAL ACCURACY	■	■	■	■	■
OVERHANGS	■	■	■		
BRIDGING	■	■	■	■	■
NEGATIVE SPACE	■	■	■	■	
RETRACTION	■	■	■		
SUPPORT MATERIAL	■	■	■	■	
SQUARENESS	■	■	■	■	■
FULL BED ACCURACY	■	■	■		
Z WOBBLE	PASS				

Price as Tested $1,599

39

- **WEBSITE** seemecnc.com
- **MANUFACTURER** SeeMeCNC
- **BUILD VOLUME** 275mm dia.×385mm tall
- **BED STYLE** Heated borosilicate glass
- **FILAMENT SIZE** 1.75mm
- **OPEN FILAMENT?** Yes
- **TEMPERATURE CONTROL?** Yes, extruder (280ºC max); bed (100ºC max)
- **PRINT UNTETHERED?** Yes, Wi-Fi (SD with optional LCD)
- **ONBOARD CONTROLS?** Yes, optional touchscreen
- **HOST/SLICER SOFTWARE** Cura
- **OS** Mac, Windows, Linux
- **FIRMWARE** Open Source - RepRapFirmware based
- **OPEN SOFTWARE?** Yes, Cura is GNU LGPL-3.0
- **OPEN HARDWARE?** Duet Wifi is open source

DREMEL 3D45

Speed and quality highlight this set-and-forget printing experience

Written by Ryan J. Priore

WHY TO BUY
The onboard camera and online Dremel DigiLab are a powerful combination for controlling and monitoring the 3D45 from anywhere you can access the web. The enclosed printer design keeps any smells contained as well as prevents dramatic ambient temperature swings during the print process.

THE 3D45 IS A POLISHED, FEATURE-RICH 3D PRINTER THAT CAN RELIABLY CRANK out beautiful prints for the makerspace, office, school, or home.

The Dremel Print Cloud (based on the 3DPrinterOS) is what differentiates this machine from others at this price point with a host of features like cloud slicing and remote monitoring/recording via the onboard camera. While last year's pre-production unit came with a custom Simplify3D profile, this year's model had Dremel's version of Cura, which didn't test as well and resulted in a lower score. This really shows how much the slicer is as important as the hardware — if not more so.

The 3D45 offers a breadth of filament options with RFID tag auto-recognition, and is aimed at educational institutions as well as makerspaces where centralized control and access is paramount. It led our pack in print speed with decent looking prints, and plowed through our test probes with ease. ◎

MACHINE RATING

	1	2	3	4	5
VERTICAL SURFACE FINISH	■	■			
HORIZONTAL FINISH	■	■	■		
DIMENSIONAL ACCURACY	■	■			
OVERHANGS	■	■	■		
BRIDGING	■	■	■	■	
NEGATIVE SPACE	■	■			
RETRACTION	■	■			
SUPPORT MATERIAL	■	■			
SQUARENESS	■	■	■	■	
FULL BED ACCURACY	■	■	■		
Z WOBBLE	PASS				

Price as Tested $1,799

35

- **WEBSITE** dremel.com
- **MANUFACTURER** Dremel
- **BUILD VOLUME** 245×152×170mm
- **BED STYLE** Heated removable glass
- **FILAMENT SIZE** 1.75mm
- **OPEN FILAMENT?** Yes
- **TEMPERATURE CONTROL?** Yes, extruder (300ºC max); bed (120ºC max)
- **PRINT UNTETHERED?** Yes, USB drive, onboard storage, and Wi-Fi
- **ONBOARD CONTROLS?** Yes, integrated LCD and SD card controller
- **HOST/SLICER SOFTWARE** DigiLab 3D Slicer software
- **OS** Mac, Windows
- **FIRMWARE** Custom
- **OPEN SOFTWARE?** Custom Cura, Cura is GNU LGPL-3.0
- **OPEN HARDWARE?** No

POWERSPEC DUPLICATOR I3 MINI

An attractive option for those looking to print on a budget
Written by Chris Yohe

Make: 2019 DIGITAL FABRICATION GUIDE — **Best Value** — Duplicator i3 Mini ★★★

THE POWERSPEC DUPLICATOR I3 MINI BY WANHAO IS EXACTLY AS IT APPEARS — a small workhorse that is both capable and affordable.

The machine ships nearly ready to run. There is no heated bed, which means that the filament choices are limited, but so are some hassles. It also has a simple scroll menu because it runs Marlin rather than fancy custom firmware. It's still a custom controller board, but the ability to modify the firmware goes a long way. The prints came out OK, albeit a bit slow for our taste with the default settings. You can easily balance quality and speed with a little tweaking of the custom version of Cura.

The i3 Mini may be a good choice for those getting started, or could serve as a bargain-basement, small-part utility machine. If you are choosing between similar printers, the open firmware of the i3 Mini might be enough to steer your decision in its direction. ⊘

WHY TO BUY
An affordable machine that lets people on a budget get started with 3D printing.

MACHINE RATING

	1	2	3	4	5
VERTICAL SURFACE FINISH	■	■	■		
HORIZONTAL FINISH					
DIMENSIONAL ACCURACY					
OVERHANGS					
BRIDGING	■	■	■		
NEGATIVE SPACE					
RETRACTION					
SUPPORT MATERIAL					
SQUARENESS		■	■	■	
FULL BED ACCURACY					
Z WOBBLE	PASS				

Price as Tested **$199**

39

- **WEBSITE** powerspec.com
- **MANUFACTURER** PowerSpec
- **BUILD VOLUME** 120×135×100mm
- **BED STYLE** Non-heated aluminum sheet with tape
- **FILAMENT SIZE** 1.75mm
- **OPEN FILAMENT?** Yes
- **TEMPERATURE CONTROL?** Yes, extruder (260°C max)
- **PRINT UNTETHERED?** Yes, microSD card
- **ONBOARD CONTROLS?** Yes, LCD with scroll wheel
- **HOST/SLICER SOFTWARE** Custom PowerSpec Cura
- **OS** Windows, Mac, Linux
- **FIRMWARE** Marlin
- **OPEN SOFTWARE?** Custom Cura, Cura is GNU LGPL-3.0
- **OPEN HARDWARE?** No

MP MINI DELUXE SLA

Produce intricate prints with this inexpensive resin machine
Written by Matt Stultz

MONOPRICE'S NEW MP MINI DELUXE SLA IS AN UPGRADED VERSION OF THE original bare-bones Wanhao D7.

The biggest change is the big touchscreen LCD, allowing you to fully control the machine — homing, moving the z-axis, testing the UV-blocking LCD, and running the prints themselves. Additionally, the z-axis has been stiffened up to help keep any z wobble lines out of the prints, and extra fans have been added to help cool the electronics and extend their life.

With a 20-micron resolution, the machine produces great prints. Details were clearly defined and parts were consistent throughout. The build space is fairly small, but that didn't really bother me as most of my SLA prints are smaller, more detailed objects to begin with. If you are in the market for a new low-cost resin-based 3D printer, the MP Mini Deluxe SLA is well worth the extra $100 to free your computer to do better things. ⊘

HIGH-RESOLUTION RESIN
Unlike fused-filament 3D printers that melt plastic to shape each layer, SLA printers produce finely detailed objects by curing a liquid resin using ultraviolet light. Want to see more of these reviewed next year? Let us know! editor@makezine.com

WHY TO BUY
As more LCD-based, low-cost resin printers hit the market, the MP Mini Deluxe SLA has a great heritage being based on the Wanhao D7 and an active community working on making it great.

PRO TIPS
Like all resin-based printers, the MP Mini Deluxe SLA can be a bit messy — make sure you keep it in an area that makes it easy to clean up after. To help keep the price down, the washout and cleaning supplies could also use a bit of help; get yourself some extra plastic bins for cleaning prints and a real scraper for getting it off the bed.

- **WEBSITE** monoprice.com
- **PRICE AS TESTED** $600
- **MANUFACTURER** Monoprice
- **BUILD VOLUME** 120×70×200mm
- **OPEN RESIN?** Yes
- **PRINT UNTETHERED?** Yes, prints can run from USB stick
- **ONBOARD CONTROLS?** Yes, touchscreen interface
- **HOST/SLICER SOFTWARE** Creation Workshop
- **OS** Windows only
- **FIRMWARE** Proprietary
- **OPEN SOFTWARE?** No
- **OPEN HARDWARE?** No

Kelly Egan

LASER CUTTERS

We're witnessing a turning point with these rapidly advancing tools

Written by Matt Stultz · Illustrated by Rob Nance

Burning for You

Hot to try out your new laser? Here's a starter project that also looks fantastic.

Kerf-Bent Wood Box by Peter Tas
Skip the steaming or hand cutting; just burn kerf cuts in the inside corners for a flawless exterior. instructables.com/id/Laser-Bending-Ply-Wood

A. Laser head
B. Lens
C. Mirrors
D. Laser tube
E. Exhaust fan
F. Gantry
G. Vacuum bed
H. Emergency shut-off
I. Controls
J. Lid

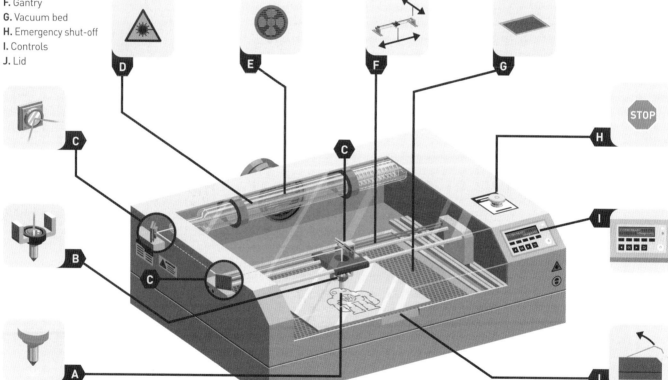

THIS YEAR, LASER CUTTERS HAVE COME INTO THEIR OWN.

Prices are declining, three of the machines we test in this issue contain camera systems, and one is even UL certified, a not-so-crazy notion for something that incinerates material to make your creations. These advancements open options for users who might not have had the patience for the lasers of the past.

FEEL THE BURN

With laser cutters it really comes down to how easy the software is to use. One name

you'll start to hear more in this space is LightBurn. Laser cutters have been long overdue for a non-proprietary, high-quality piece of control software and LightBurn is providing a solution. If you have a laser cutter with a compatible controller (an ever-growing list), this software will give you a huge upgrade to your existing machine. You can even add optical placement to your old laser with the help of an off-the-shelf webcam. We think this software is really going to catch on and may result in a growth in the number of companies producing

laser cutters and kits now that they have a reliable control option.

MARKET SEGMENTATION

When reading through these reviews, keep in mind that not all of these machines are after the same market. While the Dremel and Glowforge are out-of-the-box, high-power solutions, the Fabool is a kit, and the Emblaser 2 is a diode-based system. It's great to see more options for users — just remember to weigh them depending on your own laser-cutting needs. ◐

SAVINGS VOUCHER

	MAKE Magazine	COVER PRICE	YOU SAVE	PAY ONLY
	One Year	$59.94	33%	$39.99

BUSINESS REPLY MAIL

FIRST-CLASS MAIL   PERMIT NO. 865   NORTH HOLLYWOOD, CA

POSTAGE WILL BE PAID BY ADDRESSEE

Make:

PO BOX 17046
NORTH HOLLYWOOD CA 91615-9186

DREMEL LC40

A focus on safety makes this powerful machine a standout for educational environments *Written by Jen Schachter*

THE NEWEST ARRIVAL IN DREMEL'S DIGILAB SUITE WILL APPEAL TO BOTH new users and entrepreneurs with a streamlined interface and enough power to cut materials up to ¼" thick. The Dremel LC40 also is the first of its kind to receive a UL safety rating — an on-screen checklist precedes the start of each job, and a set of diagnostic sensors monitor the chiller, blower, exhaust, and lid, before and during the cut, protecting both user and machine.

HITS AND MISSES

An integrated water chiller and compressor, the Hex Box, comes standard at the base price and is controlled automatically. The quick-start guide covers the basics of setup: Connect a few hoses, fill the Hex Box tank, and run the wireless setup. The optics on our unit did require re-calibrating out of the box — par for the course to the experienced, but may be intimidating for new users, and is certainly beyond the "quick 20 minute setup" Dremel boasts.

The LC40's software interface is simple and intuitive — it easily imports a variety of design files, with color-mapping intact, and includes useful functions like auto array, perimeter trace, and the ability to save and run jobs from the machine memory untethered. The pre-loaded material library is indispensable, giving a starting point for settings on common materials and the option to save custom presets. DigiLab's one big oversight is its inability to cut and engrave from the same file — it reads all vector files as cut/score and all raster files as etch.

THE PROOF IS IN THE CUTTING

Once properly calibrated, the Dremel's performance was great. Cut and etch tests were clean and consistent across the bed, with a sharp kerf and minimal scorching. I hardly noticed a difference in production quality from some industrial 60W machines I have used. With basic upkeep, (and hopefully a few improvements) the Dremel will likely be a workhorse little shop laser for customizing bespoke pieces and small production runs, while the attention to safety, and adaptability with its existing DigiLab suite will be a win for educators. ⊘

WHY TO BUY
The only desktop laser with a UL safety rating, the LC40 is a good choice for educators or small-production entrepreneurs looking for a small footprint and streamlined interface.

PRO TIPS
The flex in the Dremel's case can affect alignment, so be sure you're set up on a flat surface and section of floor. (Check using a small level across the gantry and along the sides, and shim if needed so corners are at even heights.) Take the time to realign the mirrors after unboxing. It requires some patience, but will pay off with quality cuts and etches.

- **WEBSITE** digilab.dremel.com
- **MANUFACTURER** Dremel
- **PRICE AS TESTED** $5,999
- **BUILD VOLUME** Engraving: 467×305mm; Cutting: 508×305mm
- **LASER TYPE** 40W CO2 laser; Class 4 with lid open, Class 3R with lid closed; 6.5mm @ 1.2m beam size
- **CUT UNTETHERED?** Yes, can save jobs to the machine. Software runs off laser cutter itself and connects to computer through wireless connection.
- **ONBOARD CONTROLS?** Yes, touchscreen
- **SOFTWARE** Dremel DigiLab, hosted onboard
- **OS** Mac, Windows, Linux
- **FIRMWARE** Proprietary
- **OPEN SOFTWARE?** No
- **OPEN HARDWARE?** No
- **COOLING** Proprietary external water chiller/blower system called the Hex Box
- **FUME EXTRACTION** Exhaust tube to vent outside. Filtration system for indoor use can be purchased for an additional $1,000 on the base price.

I HARDLY NOTICED A DIFFERENCE IN QUALITY FROM SOME INDUSTRIAL 60W MACHINES

DREMEL
DIGILAB

GLOWFORGE BASIC

An affordable desktop machine with a refreshingly simple user experience *Written by Jen Schachter*

GLOWFORGE SALES FINALLY OPENED TO THE PUBLIC IN APRIL, AND WE CAN NOW confirm the hype. It's a capable 40W laser inside a sturdy and thoughtfully engineered enclosure, with an interface that's as easy to master as your home office printer.

MINIMAL GUESSWORK

The maximum cutting area is 495×279mm, and it can cut material up to ¼" thick. The only control on the machine is a glowing "Print" button on the case. The rest is run via Wi-Fi through the browser-based Glowforge app, including a slick automatic focus system and two onboard cameras which image the bed every time you open the lid. They also scan Glowforge's proprietary Proofgrade materials — specialized sheets of wood, acrylic, and leather, each with a coded ID sticker to automatically import their settings into your file. Non-certified materials are still safe to use, and the long-awaited custom material library should have launched by now, enabling users to save their own presets for non-Proofgrade materials.

The Glowforge is a thoroughly closed system — everything from self-contained cooling to its sealed laser head precludes the need for tweaking and tinkering. There are no auxiliary systems to monitor, and the optics stay calibrated and aligned during shipping, so you can start cutting right out of the box.

The image trace function and ready-made starter projects are a fun way to get cutting without the need for any design software. When you're ready to create your own files, you can find an extensive archive of tutorials for every function and design program on the forums. You can prep and manage files from anywhere, even with the app disconnected from the Glowforge, a blessing of the web-based software, but there is no option to hard wire to your computer.

REVOLUTIONARY TECHNOLOGY

Glowforge is an ambitious startup with a community that holds its own against the industry giants. With a truly plug-and-play package, intuitive software interface, and results comparable to machines twice the price, the Basic will appeal to just about anyone in the market for a desktop laser, from hobby users to educators. ◔

THOROUGHLY CLOSED SYSTEM PRECLUDES THE NEED FOR TWEAKING AND TINKERING

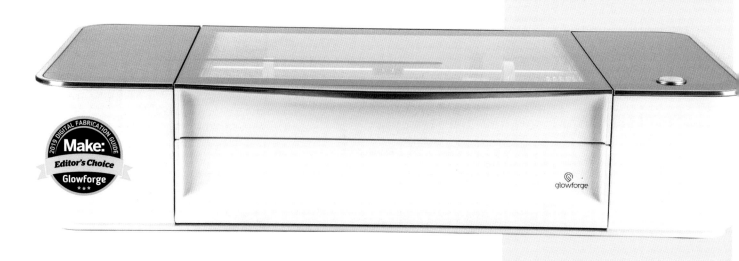

2019 DIGITAL FABRICATION GUIDE
Make:
Editor's Choice
Glowforge
★★★

glowforge

FABOOL LASER CO2

This inexpensive, modifiable kit is best for more experienced users

Written by Matt Dauray

BOASTING A GENEROUS BED SIZE, A DECENT TUBE SIZE, AND UPGRADABLE components, the Fabool 40-watt CO2 laser cutter from Smart DIYs promises a lot to potential buyers at a very reasonable price.

At its core the Fabool is a kit laser and therefore should be treated as one. The quality of the build is mostly on the builder, though we could use some better parts (*ahem, lid hinges*), and better English translation of Japanese instructions and prompts. The complexity and calibration frustrations lend themselves to more experienced users who have a thorough understanding of laser mechanics.

Although Smart DIYs software and drivers are not quite ironed out yet, the machine itself is high quality and made some decent cuts. This kit might be a great candidate for someone looking to source cheaper parts and drop a Smoothieboard or similar control board into it. Combined with software such as LightBurn, this could be a fantastic deal. 🌐

EMBLASER 2

This low-cost, super-safe diode desktop machine is a great first laser

Written by Jen Schachter

THE 5-WATT EMBLASER 2 IS SURPRISINGLY POWERFUL, AND PAIRED with full-featured software and some great safety functions, it's a viable choice for those looking to get started without the risk or expense of some CO2 machines.

Power and "arm laser" buttons are on the machine, but all other operations, including focusing, are run through open source software, LightBurn. There may be a learning curve for first-timers, but starter projects and tutorials can help get you cutting. The workflow is intuitive and has some truly clever features. The material preset library with fully customizable profiles really takes it up a notch.

Once the settings were dialed in, we were cutting ¼" ply with pretty consistent results. Where the Emblaser really shines is etching detailed raster images, making it an excellent choice for photo engraving. For those looking to add laser cutting to your tool chest or curriculum, pick up this bright offering from Australia's Darkly Labs. 🌐

Kelly Egan

CNC ROUTERS

These powerful tools can make incredible projects, big and small

Written by Matt Stultz · Illustrated by Rob Nance

Making the Cut

Put your CNC router to work with a starter project that's a keeper.

Heirloom Step Stool by Matt Stultz
Strong and beautiful, to hand down through the generations. makezine.com/go/cnc-step-stool

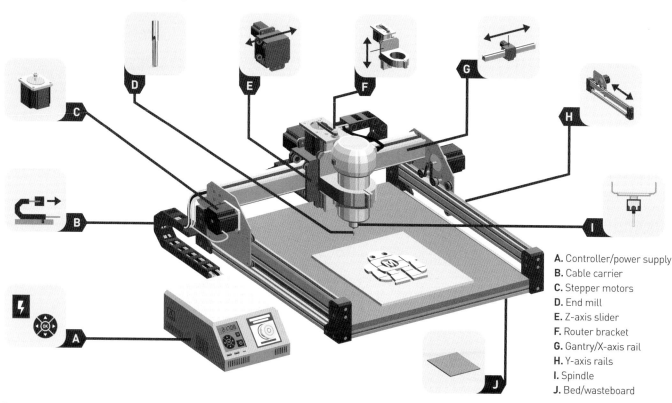

A. Controller/power supply
B. Cable carrier
C. Stepper motors
D. End mill
E. Z-axis slider
F. Router bracket
G. Gantry/X-axis rail
H. Y-axis rails
I. Spindle
J. Bed/wasteboard

THIS YEAR, WE'VE BROUGHT IN A NEW BATCH OF CNC MACHINES to review. Two of these routers, from Next Wave Automation, are available at popular woodworking stores nationwide — giving the rare option to read a CNC review and then walk into a shop and purchase it the same day. The mainstreaming of CNC machines is slowly getting underway.

MATERIAL CONSIDERATIONS

As with most benchtop CNC machines, you'll be looking at similar material capabilities with this year's selections — wood and plastics for sure, and some soft metals like aluminum if you're really careful and willing

to accept some broken end mills. If you want to get into milling steel, you'll want to look at something like the Tormach PCNC 440 (reviewed in *Make:* Volume 54).

CAM SOFTWARE

One key element in the CNC workflow is dealing with Computer Aided Manufacturing (CAM) software — the tool that converts a design into the machine commands to make the part.

Autodesk's Fusion 360 continues to expand its reach as the preferred CAM option for many manufacturers' machines — offering design and control in one package, and for free if you're a hobbyist

or student. It does, however, come with a steep learning curve that can be intimidating for new users.

Another CAM solution for makers is from Vectric. Their VCarve CAM software is especially useful for making 2D parts like furniture components or signs. For $99 per year, a verifiable makerspace will receive five full licenses of VCarve Pro Makerspace Edition, and unlimited client licenses. This allows members to set up their jobs on their own computers and save VCarve files that can then be opened on a computer running a full license to export the G-code file for their machine. We use this at Ocean State Maker Mill and it works great. ◎

OPENBUILDS MINIMILL

An accurate, customizable, and portable machine

Written by Matt Dauray

THE OPENBUILDS MINIMILL IS A SMALL WORK AREA PLATFORM DESIGNED TO be built, modified, and tinkered with. As its name implies, OpenBuilds encourages a community-based support system and provides transparency in its designs.

ALL ASSEMBLY REQUIRED

The basic bundle comes completely disassembled with only the mechanical pieces that make up the axis and a mount for a router. The only documentation is a link to a few YouTube kit assembly videos; there is no other instruction on how to mount and wire your board, motors, or accessories. Luckily, OpenBuilds provides helpful links and tips to other CNC driver boards, router options, and software recommendations.

Once you download and open Grbl Panel and connect to the board, you must correctly input all of the driver settings specific to the machine (provided on the build page), or the MiniMill will behave unpredictably — which is not ideal considering the lack of endstops. The MiniMill moves the spoilboard in the x and y directions, which can be confusing to determine which way is positive or

negative, but that can be adjusted to your personal preference either in software or by reversing the motor wiring. The knurled knobs attached to the x and y axes are a useful feature: You can turn the machine off, turn the knobs manually to position, then turn the machine on and zero. The bed size is small enough that this feature reduces setup time.

MINI BUT MIGHTY

Our test cuts produced accurate results that are a testament to the MiniMill's sturdy C-beam construction. The open air setup allows for adequate lubrication if necessary and easy cleaning. The two included handles actually make the MiniMill quite portable; we moved it around our shop multiple times with no issues.

Inherently, the gauge for a machine's usefulness is its user's intentions. As a production tool the MiniMill would fall short, but as an enthusiast's small prototyper or circuit board cutter it could be great. With this particular kit the work you put into it will determine the work you get out. And we think that's cool. ◉

WHY TO BUY

Makers with a drive to customize and a smaller operation goal will enjoy its frugality. Off the shelf parts make tailoring to a specific task more manageable.

PRO TIPS

The power supply from OpenBuilds has a voltage selector hidden away on the side of the case. Make sure it's on the right setting or at best it won't power on (at worst, it'll fry your board).

Buy endstops: they're easy enough to setup, won't break the bank, and save you from inevitable crashing on the small bed size.

The spindle/router LED ring is something you think doesn't make a difference until you don't have one. It lights up your work in 360° so that there are no shadows.

■ **WEBSITE** openbuilds.com

■ **MANUFACTURER** OpenBuilds

■ **BASE PRICE** $400

■ **PRICE AS TESTED** $820

■ **ACCESSORIES INCLUDED AT BASE PRICE** Basic bundle comes with the mechanical pieces to build the machine, no motors, board, router or power supply

■ **ADDITIONAL ACCESSORIES PROVIDED FOR TESTING** NEMA 23s, CNC xPRO V3 Controller Stepper Driver, 24V power supply bundle, DeWalt 611, spindle LED ring.

■ **BUILD VOLUME** 114×191mm cutting area, 76mm z travel

■ **WORK UNTETHERED?** No

■ **ONBOARD CONTROLS?** None, direct plug connection

■ **DESIGN SOFTWARE** SketchUp Make, anything that can post to Grbl

■ **CUTTING SOFTWARE** SketchUp SketchUcam plugin, Grbl Panel, Cut3D, MeshCAM, BobCAM

■ **OS** Default setup supports Windows only; other options support Mac and Linux

■ **FIRMWARE** Grbl, GPLv3

■ **OPEN SOFTWARE?** Yes. Grbl panel is open source. Recommended modeling and cam software is not open source. CNC driver board is fully customizable.

■ **OPEN HARDWARE?** Yes, to an extent. The kit contains a part list with part numbers, all corresponding back to OpenBuilds, but most of the parts can be sourced elsewhere. The only difficult-to-find piece is the c-beams.

THE TWO INCLUDED HANDLES ACTUALLY MAKE THE MINIMILL QUITE PORTABLE; WE MOVED IT AROUND OUR SHOP MULTIPLE TIMES WITH NO ISSUES

PIRANHA FX

Precise and compact, ideal for small shops and hobbyists

Written by Matt Dauray

PRO TIPS

Tool paths are exported individually, and the pendant G-code sender has a limit on file name length. Keep a journal of which file executes what operation.

Use double-sided duct tape to secure smaller work pieces to avoid any collisions with clamps.

Clamp pieces vertically to the edge of the table using the two holes in the front of the machine.

For a homemade kill switch, plug everything into a 15amp surge protector with a big lighted off switch and position the power strip close by.

IF YOU'RE LOOKING FOR A HEADACHE-FREE WAY TO GET INTO CNC CARVING, Next Wave Automation's Piranha Fx is a great place to start. It comes mostly assembled, and includes a license for Vetric's VCarve Desktop in the basic package, which really cuts down the learning curve.

As with most basic level setups, there are drawbacks. The Piranha does not have limit switches, and while the single lead screw saves on cost and weight, it does provide opportunity for racking under heavy stresses. Cutting area is limited to 305mm×330mm and smaller stepper motors mean feeds and speeds will be relative.

Our cut comparisons showed no obvious signs of decreased accuracy despite its smaller motors and bed size. At $1,600 this is a great machine for shops tight on space or uninterested in industrial level complexity, while still being able to produce quality parts — especially wooden signs or plaques. ◉

WHY TO BUY

Barrier of entry is super low on this machine, without sacrificing quality. Vast experience is not necessarily required and simplified control features streamline the workflow, which means less mistakes and a better user experience.

- **WEBSITE** nextwaveautomation.com
- **MANUFACTURER** Next Wave Automation
- **BASE PRICE** $1,600
- **PRICE AS TESTED** $1,700
- **ACCESSORIES INCLUDED AT BASE PRICE** 2 slotted track clamps, vee bit
- **ADDITIONAL ACCESSORIES PROVIDED FOR TESTING** Touch plate
- **BUILD VOLUME** 305×330mm cutting area, 76mm z travel
- **WORK UNTETHERED?** Yes, USB is inserted into attached pendant
- **ONBOARD CONTROLS?** Yes, power switch, attached UI pendant, no mechanical kill switch
- **DESIGN SOFTWARE** Vendor provided VCarve Desktop V8 Design. Other software will work if the Piranha post processor is used.
- **CUTTING SOFTWARE** VCarve outputs individual toolpaths to the pendant G-code sender.
- **OS** PC, Mac running conversion software
- **FIRMWARE** Proprietary
- **OPEN SOFTWARE?** No
- **OPEN HARDWARE?** No

SHARK II

A beefy frame and big motors combine for efficient production

Written by Matt Dauray

PRO TIPS

VCarve and the Shark post processor is very sensitive to node position in arc segments. It automatically slows down the cut in order to achieve the highest accuracy to your drawing. But this can be unnecessary. Keep your machine time down by reducing excessive nodes in arcs. This decreases your file sizes as well.

THE SHARK II IS THE MIDDLEWEIGHT CLASS OF CNC ROUTER BUILT BY Next Wave Automation. While not the biggest bed size of NWA's line up, it offers many of the better design features not found on their smaller machines. Larger motors, adjustable bearings, a robust HDPE/aluminum frame, and a few new software features combine for a capable and dependable user experience.

The cradle on the gantry is sized for larger 2¼ HP routers like the DeWalt 618 or dedicated spindle from NWA. The package includes a license for VCarve Desktop, and the work area is 330mm×635mm, which is enough to carve your average door plaque. In our tests, the accuracy of the Shark II was on par with the Piranha while running the same feeds and speeds, but the increased rigidity of this machine makes it great for a shop with more repetitive work — it can be run harder and faster to increase efficiency while maintaining the quality of cut. ◉

WHY TO BUY

A larger HP router cradle and beefier construction increases your capabilities and efficiencies. A larger bed size means bigger pieces, or more multiples of smaller jobs. Software upgrades and time-saving accessories make carving easier and more accurate.

- **WEBSITE** nextwaveautomation.com
- **MANUFACTURER** Next Wave Automation
- **BASE PRICE** $3,000
- **PRICE AS TESTED** $3,100
- **ACCESSORIES INCLUDED AT BASE PRICE** Router mount, control LCD
- **ADDITIONAL ACCESSORIES PROVIDED FOR TESTING** Auto zero touch plate
- **BUILD VOLUME** 330×635mm cutting area, 178mm z travel
- **WORK UNTETHERED?** Yes, USB is inserted into attached pendant
- **ONBOARD CONTROLS?** Yes, power switch, attached UI pendant, mechanical kill switch
- **DESIGN SOFTWARE** Vendor recommended and provided VCarve Desktop V8 Design. Other software will work provided the Piranha post processor is used.
- **CUTTING SOFTWARE** VCarve outputs individual toolpaths to the pendant G-code sender.
- **OS** PC, Mac running conversion software
- **FIRMWARE** Proprietary
- **OPEN SOFTWARE?** No
- **OPEN HARDWARE?** No

VINYL CUTTERS

Stickers, stencils, and so much more –
these tools are a useful addition to any workshop

Written by Mandy L. Stultz · Illustrated by Rob Nance

A. Manual feed roller
B. Cutting carriage
C. Drive belt
D. Grit rollers
E. Pinch rollers
F. Guide lines

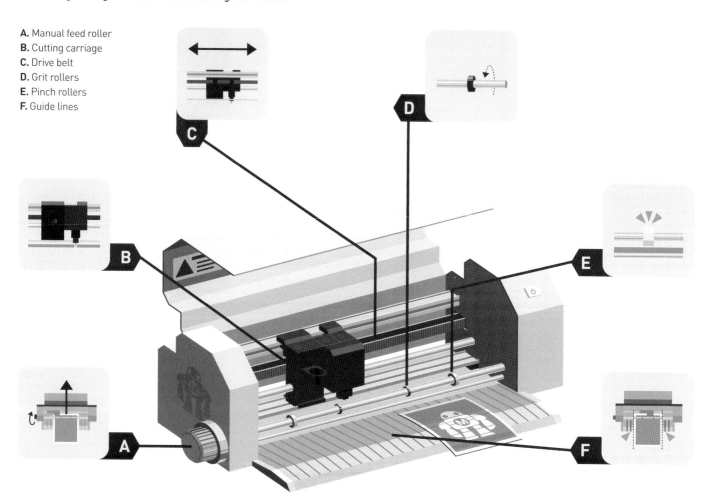

Stick and Move

A classic starter project for craft cutters, vinyl stickers make great labels and wall art. There's no limit to where they can go.

Smokey Nelson

Helmet Stickers by Becky Stern
Inspired by hand-painted helmets, Becky chose a brush script to share her thoughts about riding her motorcycle in NYC. instructables.com/id/Vinyl-Helmet-Stickers

FROM SIGNAGE TO DECOR TO CRAFTS, VINYL AND CRAFT CUTTERS ARE THE UNSUNG HEROES OF DIGIFAB.

Professional-grade vinyl cutters continue to enjoy solid improvements. Machines are moving to streamlined designs with more intuitive interfaces, and software features are expanding beyond the bare-bones offerings of the past.

But it's craft cutters that have seen the most dramatic changes recently. Moving far beyond the small machines with limited onboard controls and sticky-mats of just a few years ago, we're getting touchscreens, scanning capabilities, and new adapters to cut vinyl from continuous rolls. Self-adjusting blades now help to eliminate any guesswork, and machines are more capable than ever of working with a wide variety of materials — adding options like leather and fabric to the mix. And the interfaces for craft-oriented cutters continue to shine, as most are focused on ease of use. Whether installed or in the cloud, you can learn the basics for nearly all of them in a few hours. ◐

SILHOUETTE CAMEO 3

Size and versatility combine for an impressive machine

Written by Mandy L. Stultz

THE SILHOUETTE CAMEO 3 IS A POWERHOUSE — the wide range of materials it can handle allows it to go above and beyond scrapbooking, becoming a capable tool for other crafters as well.

The new AutoBlade feature presets were fairly accurate, but it did exhibit some drag out/catch with super fine lines on regular vinyl and foil vinyl. The AutoBlade settings aren't stored by the machine, which is somewhat annoying as that requires going through its dial-in process for a test cut as well as the final. Also, I must warn you that this feature's motions seem somewhat severe and it sounds as if the machine is malfunctioning; rest assured, it is not.

You may need to find your way around some of its quirks, but all in all, the Cameo 3 is a powerful machine, great for the avid crafter who has room for its significant footprint, or makerspaces looking for a multifunctional tool. ⊘

- **WEBSITE** silhouetteamerica.com
- **MANUFACTURER** Silhouette
- **PRICE AS TESTED** $300
- **CUTTING SIZE** 305×305mm, but able to cut 305×3048mm via roll feeder
- **CUT UNTETHERED?** Yes, Bluetooth or Bluetooth compatible depending on version; USB key
- **ONBOARD CONTROLS?** Yes, touchscreen to control machine
- **CONTROL SOFTWARE** Silhouette Studio
- **OS** Windows Vista, Windows 7-10, MAC OSX 10.7 or later
- **OPEN SOFTWARE?** No

SILHOUETTE PORTRAIT 2

This portable yet powerful cutter is perfect for crafters

Written by Mandy L. Stultz

WHILE ONE OF THE SMALLER CUTTERS ON THE MARKET, THE SILHOUETTE Portrait 2 is a mighty machine for its size, capable of cutting 100 different materials.

For portability and stowability, this machine can't be beat. Around 3.5lbs and about 12"×5" dimensionally, it fits easily into a tote for vinyl cutting on the go, and could be stored in a drawer or cabinet when not in use. Pens and markers allow you to use it as a plotter drawing and lettering tool too. Onboard Bluetooth means no extra cable.

AutoBlade is the newest feature — loud and with some quirks, as mentioned in the Cameo 3 review above, but ultimately does seem to work well.

The Portrait 2 is ideally for the individual crafter, and while it seems sturdy enough to handle what's thrown at it, makerspaces or multiple-use situations would likely want a larger, more robust machine. For what it's worth, an original Portrait has survived several years at my hackerspace, but it is not our primary or most heavily used cutter. ⊘

- **WEBSITE** silhouetteamerica.com
- **MANUFACTURER** Silhouette
- **PRICE AS TESTED** $200
- **CUTTING SIZE** 203×305mm but able to cut 203×3048mm via roll feeder
- **CUT UNTETHERED?** Yes, Bluetooth included or Bluetooth compatible, depending on version
- **ONBOARD CONTROLS?** Yes, load/unload buttons, pause, power, and Bluetooth button
- **CONTROL SOFTWARE** Silhouette Studio
- **OS** Windows Vista, Windows 7-10, Mac OSX 10.7 or later
- **OPEN SOFTWARE?** No

Kelly Egan

Want to get into the nitty-gritty of machine comparison? We've pulled all the data and specs from our reviews this year and past to create the following tables — dig through for help finding the exact machine for your needs, or just to geek out. Find even more review info online at makezine.com/go/digifab-2019.

FFF PRINTER TEST SCORES

Legend:
- Vertical Surface Finish
- Horizontal Surface Finish
- Dimensional Accuracy
- Overhangs
- Bridging
- Negative Space
- Retraction Performance
- Support Material
- Squareness
- Full Bed Accuracy
- Z Wobble

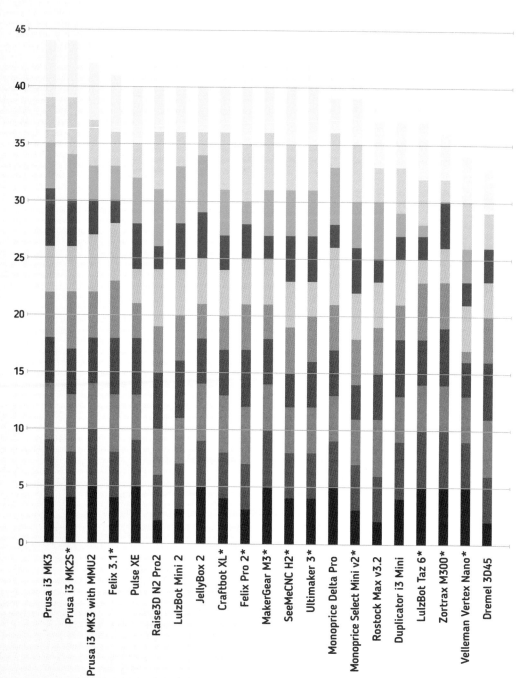

* Find these reviews online and in previous issues of Make:.

FFF COMPARISON

Machine	Manufacturer	Price	Build Volume	Open Filament	Heated Bed	Wi-Fi	Open Source	Review
Craftbot XL *	Craft Unique	$1,899	300×200×440mm	✓	✓	✓		Vol. 60
Delta Pro	Monoprice	$1,499	270mm dia.× 300mm	✓	✓	✓		page 29
Dremel 3D45	Dremel	$1,799	245×152×170mm	✓	✓	✓		page 30
Duplicator i3 Mini	PowerSpec	$199	120×135×100mm	✓				page 31
Felix 3.1 *	Felix Printers	$1,620	255×205×225mm	✓	✓			Vol. 60
Felix Pro 2 *	Felix Printers	$2,708	237×244×235mm	✓	✓			Vol. 60
Hacker H2 *	SeeMeCNC	$549	175(dia.)×200 mm OR 140(dia.)×295mm	✓			✓	Vol. 60
JellyBox 2	Imade3D	$949 (kit)	170×160×145mm	✓			✓	page 27
LulzBot Mini *	LulzBot	$1,250	152×152×158mm	✓	✓		✓	Vol. 60
LulzBot Mini 2	LulzBot	$1,500	160×160×180mm	✓	✓		✓	page 29
M3 *	MakerGear	$2,550	203×254×203mm	✓	✓	✓	✓	Vol. 60
M300 *	Zortrax	$2,990	300×300×300mm	✓	✓			Vol. 60
MakeIt Pro-L *	MakeIt	$3,199	305×254×330mm	✓	✓			Vol. 60
N2 *	Raise3D	$2,999	305×305×305mm	✓	✓	✓		Vol. 60
N2 Pro2	Raise3D	$3,999	305×305×300mm	✓	✓	✓		page 26
Prusa i3 MK2S *	Prusa Research	$899, $599 (kit)	250×210×200mm	✓	✓		✓	Vol. 60
Prusa i3 MK3	Prusa Research	$999	250×210×210mm	✓	✓		✓	page 25
Prusa i3 MK3 with MMU2	Prusa Research	$299 (add-on) + $999 (machine) = $1,298	250×210×210mm	✓	✓		✓	page 26
Pulse XE	MatterHackers	$1,595	250×220×215mm	✓	✓	✓	✓	page 28
Replicator+ *	MakerBot	$2,499	295×195×165mm			✓		Vol. 57
Rostock Max v3.2	SeeMeCNC	$1,599	275mm dia. ×385mm	✓	✓	✓	✓	page 30
Select Mini v2 *	Monoprice	$189	120×120×120mm	✓	✓	✓		Vol. 60
Taz 6 *	LulzBot	$2,500	280×280×250mm	✓	✓		✓	Vol. 60
Ultimaker 2 Extended+ *	Ultimaker	$2,999	223×223×304mm	✓	✓		✓	Vol. 54
Ultimaker 3 *	Ultimaker	$3,495	176×182×200mm	✓	✓	✓	✓	Vol. 60
Vertex Nano *	Velleman	$349 (kit)	80×80×75mm	✓				Vol. 60

SLA COMPARISON

Machine	Manufacturer	Price	Build Volume	Style	Open Resin	Print Untethered	Review
DLP Pro+ *	mUVe 3D	$1,899	175×99×250mm	DLP	✓	✓	Vol. 54
Duplicator 7 *	Wanhao	$495	120×68×200mm	DLP	✓		Vol. 60
Form 2 *	Formlabs	$3,350	145×145×175mm	SLA	✓	✓	Vol. 48
Moai *	Peopoly	$1,295	130×130×180mm	SLA	✓	✓	Vol. 60
MP Mini Deluxe SLA	Monoprice	$600	120×70×200mm	SLA	✓	✓	page 31
Nobel 1.0 *	XYZprinting	$700	128×128×200mm	SLA		✓	Vol. 60

CNC COMPARISON

Machine	Manufacturer	Base Price	Price as Tested	Build Volume	CAM Software	Work Untethered	Review
Asteroid *	Probotix	$3,649	$4,178	635×940×127mm	Vectric Cut2D, Vectric PhotoVCarve, MeshCAM, Cut3D, VCarve Pro		Vol. 60
Benchtop Pro *	CNC Router Parts	$3,250	$4,062	635×635×171mm	Fusion 360		Vol. 60
Handibot 2 *	ShopBot	$3,195	$3,195	152×203×76mm	VCarve Pro ShopBot Edition	Yes, wireless	Vol. 60
High-Z S400T *	CNC-Step	$5,299	$14,730	400×300×110mm	Any CAD package should work		Vol. 60
MiniMill	OpenBuilds	$400	$820	114×191×76mm	SketchUp SketchUcam plugin, Grbl Panel, Cut3D, MeshCAM, BobCAM		page 38
Piranha Fx	Next Wave Automation	$1,600	$1,700	305×330×76mm	VCarve outputs individual toolpaths to the pendant G-code sender	Yes, USB is inserted into attached pendant	page 39
Shapeoko XXL *	Carbide 3D	$1,730	$1,730	838×838×76mm	Carbide Create or MeshCAM		Vol. 54
Shark II	Next Wave Automation	$3,000	$3,100	330×635×178mm	VCarve outputs individual toolpaths to the pendant G-code sender	Yes, USB is inserted into attached pendant	page 39
ShopBot Desktop Max *	ShopBot	$9,090	$9,285	965×635×140mm	VCarve Pro		Vol. 54
Sienci Mill One Kit V2 *	Sienci Labs	$399	$498	235×185×100mm	Universal GCode Sender, any G-code sending software		Vol. 60
X-Carve *	Inventables	$1,329	$1,493	750×750×67mm	Easel		Vol. 54

LASER COMPARISON

Machine	Manufacturer	Price	Cutting Size	Control Software	Review
Dremel LC40	Dremel	$5,999	Engraving: 467×305mm; Cutting: 508×305mm	DigiLab	page 33
Emblaser 2	Darkly Labs	$2,495	500×300mm	LightBurn	page 35
Fabool CO2	Smart DIYs	$2,698	600×440mm	Fabool Desktop	page 35
Glowforge Basic	Glowforge	$2,495	495×279mm	Glowforge	page 34
Muse *	Full Spectrum Laser	$5,000	508×305mm	RetinaEngrave v2	Vol. 60
Voccell DLS *	Voccell	$4,999	545×349mm	Vlaser	Vol. 54

WATERJET COMPARISON

Machine	Manufacturer	Price	Maximum Cutting Dimensions	Control Software	PSI	Review
ProtoMax	Omax	$19,950	305×305mm	Proto Layout, Proto Make	30,000PSI	page 23
Wazer	Wazer	$4,499	305×457mm	WAM (Wazer CAM)	Undisclosed	page 22

VINYL CUTTER COMPARISON

Machine	Manufacturer	Price	Cutting Size	Cut Untethered	Control Software	Review
Cameo 3	Silhouette	$300	305×305mm (305×3048mm w/roll feeder)	Yes, Bluetooth included or Bluetooth compatible (depending on version), USB key	Silhouette Studio	page 41
Curio *	Silhouette	$250	216×152mm		Silhouette Studio	Vol. 60
Portrait 2	Silhouette	$200	203×305mm (203×3048mm w/roll feeder)	Yes. Bluetooth included or Bluetooth compatible (depending on version)	Silhouette Studio	page 41
ScanNCut2 CM350 *	Brother	$299	298×298mm	Yes, with purchase of the ScanNCut online activation card	CanvasWorkspace	Vol. 60
Titan 2 *	USCutter	$995	609×7620mm		VinylMaster Cut OEM (PC); Sure Cuts a Lot Pro (Mac)	Vol. 60

LED "Nixie" Display

Laser-cut and edge-lit, these 10-digit numeric displays are
bigger, brighter, and safer than the old tubes

Written by Florian Schäffer

NEO RETRO: We built this "Lixie" clock with just two
digits, so it displays the hours and minutes alternately. The
number 3 on the left is engraved as a single line, imitating
the appearance of the original Nixie tube displays.

TIME REQUIRED:
3–4 Hours

DIFFICULTY:
Intermediate

COST:
$30 per digit, plus Arduino

MATERIALS
» **Arduino Uno or Nano microcontroller board**
FOR EACH "DIGIT":
» **LED Nixie printed circuit board (PCB)** Download the design files (*.sch* or *.brd*) free from makezine.com/go/led-nixie and send them to OSH Park or a similar PCB maker.
» **RGB LEDs, WS2812B type, 5mm×5mm (20)** such as Adafruit #1655
» **Capacitors, 100nF (0.1µF), SMD 1206 package (20)**
» **Right angle female headers, 1×4 pin, 0.1" spacing**
» **Right angle male headers, 1×4 pin, 0.1" spacing**
» **Resistor, 220Ω**
» **Clear acrylic sheet, 3mm (⅛"), extruded (XT) type, about 0.5m² (~775 sq. in.)**
» **M3×20mm brass screws with nuts (4)**
» **Nonconductive foam, about 8cm×8cm (3"×3")** for light grid

OPTIONAL:
» **Real-time clock (RTC) module, DS3231 type** if you're making a clock
» **5V DC power supply** if you're driving more than two LED Nixie digits

TOOLS
» **Laser cutter**
» **Soldering iron and solder**
» **Computer with Arduino IDE** free download from arduino.cc/downloads
» **Project code and templates** Free download from the project page makezine.com/go/led-nixie, includes laser cutting templates, PCB files, and two Arduino sketches: one to drive your LED Nixies and another to use them in a two-digit clock.

Originally published in Make: German edition, issue 4/2018. Translation by Niq Oltman.

FLORIAN SCHÄFFER has been slinging a soldering iron since his youth. For the past two years he's been working as an editor for the German edition of *Make:*.

Our LED-based replica doesn't quite match the charm of a real Nixie tube, but is much easier to make, and more useful.

The legendary Nixie tubes are retro and cool — little glowing neon lamps that display numbers or symbols (Figure Ⓐ). Too bad they're no longer mass-produced, and the few that are still being made are quite expensive. In addition, they require sophisticated, high-voltage driver circuits. The good news: using multicolor LEDs and acrylic plates engraved with a laser cutter, you can re-create the Nixies' retro charm, scale them up, and customize the design to your taste. In this project we'll show you how to make a great-looking "LED Nixie" numeric display with illuminated digits, driven by an Arduino at a safe, cool 5V.

The working principle of the Nixie replica is simple, but effective: Ten transparent plates made out of extruded acrylic sheet (*polymethyl methacrylate* or PMMA, also known as Lucite, perspex, or plexiglass) are engraved with the numerals 0 to 9. Stacked on top of one another, each plate is illuminated from the edge using multicolor LEDs. Due to the internal refraction of light within the acrylic plate, only the numeral that's selected will actually light up, while the plates' surfaces, and all the other numerals in the stack, will remain transparent and colorless.

We didn't come up with this idea — we've seen it demonstrated at Maker Faires in the U.S. and Europe. Connor Nishijima, a maker in Utah, coined the name "Lixie" for his projects, inspiring many other makers.

Our own design aims to show you how easy it is to build LED Nixies and make your own creations. You'll need only basic Arduino skills, plus some soldering ability to populate the bare PCB.

LASER CUTTING
The cutting templates include 10 numeral plates and a clamping plate (single- or double-digit) to hold them all together. You'll cut the red lines and etch the green lines (numerals). We chose two typefaces that we think work well. The thin type reminds us of the classic Nixie tubes, while the double-stroke type is more readable in brightly lit environments but conceals a bit more of the numerals further back in the stack.

We're building a clock with our LED Nixies, so we need numbers, but there's no limit to the type of letters, symbols, or designs you can use. Say you want to make a weather station that fetches its data over Wi-Fi. Your temperature display could be enhanced using a color range (say, blue to red), and you could use custom symbols for rain, clouds, sunshine, and thunderstorms.

ELECTRONICS
Pretty much all LED Nixies are based on WS2812B-type RGB LEDs, or the equivalent SK6812. These parts are popular as they can be controlled easily using very little hardware and code. Make sure to look for the "B" in WS2812B — this is a redesigned part with only four instead of six connectors on the SMD package. Adafruit offers various small boards using these LEDs in their NeoPixel line, and they provide a software

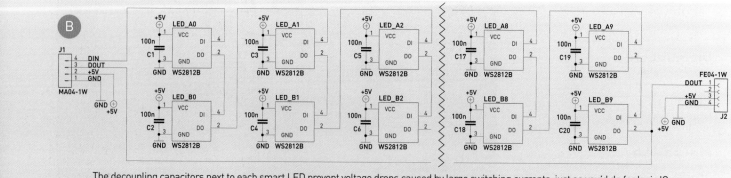

The decoupling capacitors next to each smart LED prevent voltage drops caused by large switching currents, just as you'd do for logic ICs.

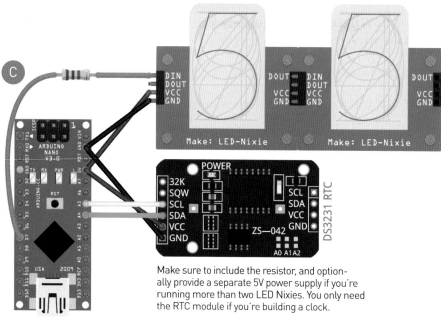

Make sure to include the resistor, and optionally provide a separate 5V power supply if you're running more than two LED Nixies. You only need the RTC module if you're building a clock.

BUILDING AN LED NIXIE

The light from the LEDs on the circuit board shines into the tabs on the acrylic plate, which is engraved with a numeral. The foam mask on the board prevents the light from radiating sideways and straying into adjacent plates.

Because the light enters the acrylic from the edge, total internal reflection occurs, and the acrylic behaves like a *light pipe*. The light's path is broken only at the engraved lines (and at the edges of the plate) so that only these spots light up — and the digit shines.

Numeral plate

Clamping plate

Foam mask

LED   LED

Circuit board

library to control them from an Arduino.

While the Lixies we found on Tindie only have a single 1,000µF electrolytic capacitor in parallel to the LED power supply, we stuck to the datasheet (Figure B) and gave each LED its own 100nF (0.1µF) cap, as these parts aren't just simple LEDs but also contain logic circuits. Adafruit's NeoPixel Überguide (learn.adafruit.com/adafruit-neopixel-uberguide) gives an explanation of how the WS2812B LEDs work, and why you should absolutely include a 220Ω–500Ω resistor on the data line between your microcontroller and the first LED (Figure C). This resistor is not part of the circuit board, so you need to wire it to the data line.

To light up the numerals evenly, we use two RGB LEDs for each, spaced apart appropriately on the board: two LEDs for the 0, another two for the 1, and so on. This setup differs from the Lixie design, which is why the code is incompatible. (While each LED can be controlled individually, we typically use the same color for the two LEDs on each numeral.) The 20 LEDs comprising a Nixie "digit" form a unit, with a connector on the side where the next unit may be attached. Using this kind of cascade, you can easily connect 500 LEDs (25 LED Nixies), possibly even more.

The current drawn by a large number of LEDs may pose a problem, however. Each RGB LED may require up to 60mA if all three color components are set to maximum intensity. This adds up to 1.2A for a single LED Nixie — way too much for the outputs on an Arduino. But we'll typically only light up one symbol per Nixie, so only two LEDs will be on at a time, drawing up to 120mA. Under these conditions, an Arduino can drive two LED Nixies. If your display needs more digits than that, you'll have to provide a separate 5V power supply with sufficient output current.

MECHANICAL CONSTRUCTION

The male and female header connectors on the side of each LED Nixie are arranged so that they can easily be joined into a row. The Data Out pin from the last LED is wired both to the male "input" header and the female "output" header, so that it will connect to the input of the next Nixie. If you're building a display with more than a few Nixies, you can improve its mechanical stability by attaching them all to a backplate.

As the LEDs' beam angle is too wide for our thin acrylic plates, we need to focus them using a light grid made of foam (Figure D). The tabs on the numeral plates fit into the openings in the grid. They're all held down by a clamping plate using four M3 screws; be careful not to let the tabs press directly on the sensitive LEDs.

In our first attempts, we used anti-static ESD foam that we shaped using a laser cutter. (The template for cutting the foam is included in the downloads.) It took us quite a while to notice that this type of foam was making our prototype behave erratically — the signals controlling the LEDs were so delicate that even barely touching the conductive foam would cause malfunctions. We now use plain (non-conductive) foam, and our device works fine (Figure E).

ARDUINO DRIVER

To drive the LED Nixies, we used the *Adafruit_NeoPixel.h* library, which you can download and install using the Arduino IDE's library manager. This library provides everything you need to drive chained WS2812B-type LEDs individually. When initializing a new instance, the constructor (see callout **1** on the code listing, on the following page) is called with three arguments: the number of connected LEDs, the number of the I/O pin to use as data input, and what clock speed to use for data transmission. We added two functions that do all the work of controlling the individual LEDs to display a number on a chain of Nixie units, so you don't have to.

The example code in *led-nixie.ino* shows how easy it is to display numbers. Two chained LED Nixies are connected to an Arduino. The **DIGITS** constant **2** must be set to the number of chained Nixie units you're using. It's used by the driver setup code as well as the code handling the calculations for the number display. The control **PIN** to be used is set in **3**

For each display unit ("digit"), you need a PCB, a light grid, 10 acrylic plates engraved with numerals, and a clamping plate to hold everything in place. The parts are joined using four screws.

Here's our completed two-digit "Lixie" clock, with Arduino and RTC module.

Florian Schäffer, Martina Bruns

and may be chosen arbitrarily (the default is pin 7). As soon as the code has been compiled and flashed onto the Arduino, you'll see a looping display counting up from zero to 99, using random colors.

The **OutNumber()** function takes three arguments — the number to display, color to use, and a flag that controls whether leading zeroes should be shown — then splits the number into individual digits and sends them to the output, while checking how many digits can be displayed at all.

In our schematic in Figure C, we included a DS3231 real-time clock (RTC) module. Adafruit's tutorial at learn.adafruit.com/adafruit-ds3231-precision-rtc-breakout describes the use of this module and why it's superior to the DS1307. To set the clock, you'll hard-code the current time into *led-nixie_clock.ino*, compile, and flash. The Arduino will pass the time to the RTC, which stores it locally thereafter. You'll then want to comment out the part of the code that sets the clock, so the RTC won't be reset to the old time on each reboot. Your two-digit clock will display hours and minutes alternately, using changing colors.

CHOPPING UP NUMBERS

When displaying numbers, you'll often encounter the problem that a lot of displays can't directly take multi-digit numbers as input. The number needs to be split apart, so that each individual display will light up with the respective digit value for each position (ones, tens, hundreds, and so on). To achieve this, we'll use a simple trick involving division.

Let's chop up a three-digit number into its three digits: hundreds, tens, and ones. First we divide the original value by 100 ❹ and then divide the result by 10 to find the *remainder* (this is known as a *modulo* operation, using the **%** operator) ❺. So, to split up the number 201, division by 100 gives 2.01, and this modulo 10 gives **2**, the last digit before the decimal point, and the right value for the hundreds. This result is passed onto the output routine ❻. In the next step (tens), the original number is divided by 10 ❼ (20.1), and the result of this modulo 10 ❽ gives us **0** for the tens. To get the ones, the original number is just taken modulo 10 ❾, which results in the digit value **1**. (We've used decimals here for illustrative purposes; the Arduino actually does the math with whole integers only.)

```
#include <Adafruit_NeoPixel.h>

uint8_t PIN = 7; ❸
uint8_t DIGITS = 2; ❷
Adafruit_NeoPixel pixels = Adafruit_NeoPixel(DIGITS * 20, PIN, NEO_GRB + NEO_
KHZ800); ❶

struct RGB {
  uint8_t red;
  uint8_t grn;
  uint8_t blu;
};
                              ❻
void OutDigit(uint8_t digit, uint8_t value, struct RGB* color) {
  pixels.setPixelColor((digit * 20) + (value * 2), pixels.Color(color->red,
  color->grn, color->blu));
  pixels.setPixelColor((digit * 20) + (value * 2) + 1, pixels.Color(color-
  >red, color->grn, color->blu));
}

void OutNumber(uint16_t value, struct RGB* color, uint8_t leadingZero) {
  uint8_t pos = 0;
  pixels.clear();   // clear all color values

  // Ten-thousands
  if (((value >= 10000) || leadingZero) && DIGITS >= 5) {
    OutDigit(pos, (value / 10000) % 10, color);
    pos++;
  }
  // Thousands
  if (((value >= 1000) || leadingZero) && DIGITS >= 4) {
    OutDigit(pos, (value / 1000) % 10, color);
    pos++;
  }
  // Hundreds
  if (((value >= 100) || leadingZero) && DIGITS >= 3) {
    OutDigit(pos, (value / 100) % 10, color);
    pos++;        ❹        ❺
  }
  // Tens
  if (((value >= 10) || leadingZero) && DIGITS >= 2) {
    OutDigit(pos, (value / 10) % 10, color);
    pos++;        ❼        ❽
  }
  // Ones
  OutDigit(pos, value % 10, color);
                    ❾
  pixels.show();
}

void setup() {
}

void loop() {
  RGB color;
  uint8_t u, i;
  pixels.begin();

  while (1) {
    for (u = 0; u <= 9; u++) {
      color.red = random(0, 25) * 10;
      color.grn = random(0, 25) * 10;
      color.blu = random(0, 25) * 10;
      for (i = 0; i <= 9; i++) {
        OutNumber(u * 10 + i, &color, 1);
        delay(100);
      }
    }
  }
}
```

PLAY WITH YOUR DIGITAL DISPLAY

Your LED Nixies are ready for the world! Use them for clocks and countdowns, hit counters and scoreboards, gauges and meters of all kinds — or whatever idea you've got. And if you're not keen on building them from scratch, you can order them ready-made from makers like Connor Nishijima (tindie.com/stores/connornishijima) and LED-Genial (led-genial.de/LED-Nixie) and incorporate them into your projects that way.

We look forward to hearing your ideas for using LED Nixies, and seeing your designs for custom display symbols, especially very small or very large ones! Share your projects at makershare.com, and then let us know in the comments on the project page at makezine.com/go/led-nixie. ◗

Written by Alex Glow

Archimedes: **AI Robot Owl**

Print and wear a superb owl that uses AI vision to see faces, judge emotions, and respond

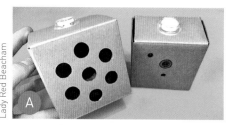

Lady Red Beacham

A

ALEX GLOW creates video and projects as the lead Hardware Nerd at Hackster (hackster.io). She was a FIRST Robotics kid, and later director of AHA and Noisebridge hackerspaces. Her favorite projects include brain-controlled wings, a song that orbited the Earth, and her robotic owl familiar.

Meet my familiar, Archimedes! He's a robotic owl who sits on my shoulder and detects the emotions of people around me. Then he gives feedback via colored lights and little beep-boops. He can move and look around, too — thanks to a pan/tilt servo gimbal with an Arduino controlling the motors.

THE BRAIN

The first part of the build is assembling the Google AIY Vision kit. This "DIY AI" introductory kit forms Archimedes' brain and sensory system: a Raspberry Pi Zero

W with "bonnet" add-on, camera, piezo speaker, and multicolor light-up button. Its fold-together cardboard enclosure is really fun to assemble (Figure A).

Just follow Google's official instructions to put the kit together, and test the setup to make sure it works. You should be able to run the default Joy Detector demo without any extra coding, but of course, you can tweak it if you want — it's written in Python. The Joy Detector uses machine learning to detect if a person is smiling (the button turns yellow) or frowning (the button turns blue). For really big expressions, the buzzer

B

To USB cable

TIME REQUIRED:
20-30 Hours

DIFFICULTY:
Advanced

COST:
$150–$180

MATERIALS
» **Google AIY Vision Kit** includes a Raspberry Pi Zero Wireless computer
» **Servomotors, metal gear, high torque (2)** I used Tower Pro MG996R.
» **Pan/tilt assembly for servos** such as RobotShop #RB-Lyn-101 and RB-Lyn-81, robotshop.com
» **Arduino microcontroller board** I used a MKR1000.
» **3D-printed parts: head, top hat, right wing, left wing, and front feather accent** Download the free 3D files at bit.ly/robotowl.
» **Sacrificial USB cables**
» **USB 5V power supplies** You'll need 3 ports total.
» **Double-sided foam mounting tape**
» **Aluminum armature wire**
» **Hookup wire**
» **Heat-shrink tubing**
» **Bicycle inner tube**
» **Old CD**
» **Scarf or bandana**

TOOLS
» **Computer with Arduino IDE** free download from arduino.cc/downloads
» **3D printer (optional)**
» **Soldering gun and solder**
» **Hot glue gun**
» **Heat gun or lighter**
» **Needlenose pliers**
» **Scissors**
» **Electrical tape (optional)**

will sound too. If the camera sees more than one face, it adds up their joy scores.

TIP: Make sure you install the plastic standoffs to protect the Pi and bonnet from mechanical stress. Mine broke when I took the kit apart later, so maybe get extra standoffs.

BODY MOVEMENTS

Two servomotors allow Archimedes to tilt his body up and down and rotate his head side to side. I started with a pan/tilt mechanism from HackerBoxes' "Vision Quest" kit (hackerboxes.com/products/hackerbox-0024-vision-quest) but you can use any pan/tilt gimbal setup; there are 3D-printed parts available on Thingiverse, if you already have a couple of servos lying around. Make sure the "tilt" part is on the bottom and the "pan" on top — since owls don't usually pivot on their legs.

Do a quick calibration on your servos and check their tolerances. What's the furthest they can go in either direction? (If they start humming or buzzing, they're pushed too far.) Edit the Arduino sketch to match that; you can download the code from bit.ly/robotowl. Right now the Arduino just moves the servos randomly so Archimedes can look around for faces. But you can modify the code for different behavior or, even better, figure out how to make the Raspberry Pi do it!

I do recommend servos with metal gears; they're a bit heavy, but they can definitely handle controlling a robot like this. If you're using 3D-printed mounts instead of metal mounts, you might be able to get away with plastic-gear servos. Plus, the whole robot would be a bit lighter, which might be nice.

The two servos are connected to the Arduino as shown in Figure **B**.

Take the time to solder everything together (Figure **C**), and once everything's working, you can cover it in hot glue to insulate and stabilize the connections. I used a small protoboard from another project to make the branching connections, to ensure that nothing would short out.

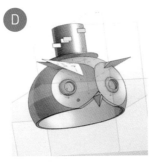

3D PRINTING

I used Onshape, a browser-based CAD modeling tool, to design Archimedes' 3D-printed body parts. He comes in five pieces: the head (Figure **D**), top hat (Figure **E**), left and right wings, and front feather accent (Figure **F**). The STL files are available from my write-up on Hackster (bit.ly/robotowl). You can also use a service

Alex Glow

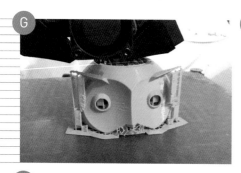

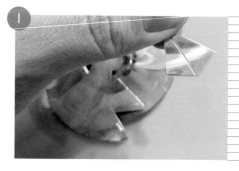

like 3D Hubs or Shapeways if you don't have your own 3D printer.

Want to mod the design? Please do! The main considerations are:

» The "brain" electronics must fit inside the head (the cables are long, but not long enough to reach down below the servo gimbal).

» Make sure you can easily accommodate the piezo buzzer and camera modules — especially the camera. If you resize the models, they'll need some extra tweaking.

My owl head came out a bit funny when I printed it (Figures **G** and **H**), because the beak wasn't well supported. So I used an old CD to give him a shiny cybernetic beak. It looks great! To do this, put the CD in a bowl and cover it in boiling water. Let it sit for 5 minutes, and you'll notice that it has delaminated a bit – the two layers have started to split apart, with the shiny memory foil on one side. Split it the rest of the way and you can cut it apart with scissors to form your preferred beak shape; I did it in two symmetrical pieces (Figure **I**). Then hot-glue it to his face. Put the bare plastic side facing out. This is surprisingly durable, but keep the CD on hand for repairs.

ROBOT ASSEMBLY

Once you're ready to put the whole robot together, grab your 3D-printed head and top hat. Install the button in the top hat, threading the cables down through the hole, and seal it in place with hot glue. Then, glue the hat to the top of the head (Figure **J**).

The camera looks out through one eye, and the piezo buzzer through the other; I've hot-glued them in place such that the glue is hidden. They're quite sturdily attached (I tend to get enthusiastic about the glue). The status LED just kind of dangles down; I think it looks kind of cool that way, and it hasn't caught on anything yet, but you might want to secure yours in place.

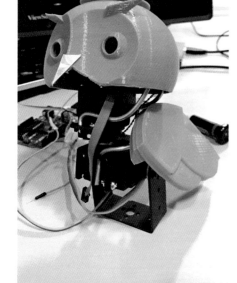

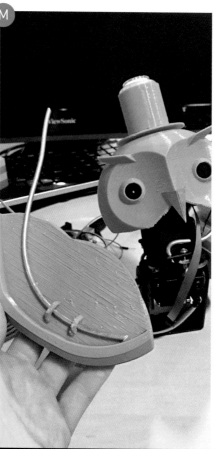

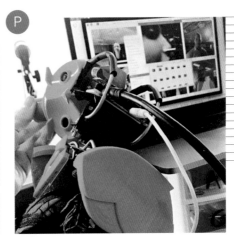

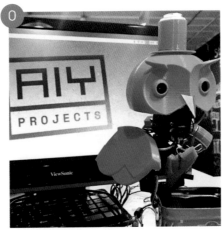

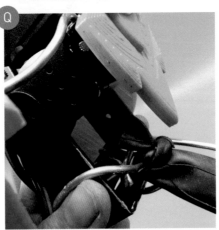

Grab your armature wire — this is a sturdy wire, typically used for supporting sculptures, which can be found at art stores. Find a good spot on the top servo mount to attach a couple of loops of it, which will support the head. Once you've done that, slide heat-shrink over the wire so it doesn't short out the electronics (Figure K).

Next, if your servo mounts are made of metal, put some electrical tape on top to insulate them from the Pi. Then put a piece of double-sided foam tape on top, and put the Pi on top of that. This will hold it in place pretty sturdily. Place the head over the top of the electronics (Figure L).

Now, use two pieces of armature wire to attach the wings (Figures M and N). Loop one end through a hole in your bottom servo mount, thread the wing onto it, and then loop the wire back through another hole in the servo mount. The wire thickness should be compatible with the wing loops, and they should be pretty sturdy; just don't apply too much torque, and they should work fine for months to come. Turn the robot's head to make sure it doesn't crash into the wings, causing damage and unhappy servos. Check the extremes in your Arduino sketch, too.

NOTE: Servomotors don't like to be turned too much by hand, since it puts force on the gears. An occasional check is fine, though.

Now for the final 3D-printed piece: Attach the feather puff loosely to the front of the bottom servo (Figure O). I used a loop of thin wire, fashioned into a bow tie.

You might want to secure the head to its wire loops, since it can flop around … but that also makes it harder to show off the brain to curious friends (Figure P).

SHOULDER MOUNT
Finally, it's time to rig your robot for wearability!

Take a piece of armature wire about 6 feet long (or a bit more). Wrap one end around and through the base of the servo mount, so that it supports the servo stably. In the future, I plan to model a custom 3D-printable shoulder mount, because the wire can get loose and wiggle around, but it'll do for now, and a robot with lighter servos/supports might have better luck.

Double up the remainder of your armature wire. This is where the bike

inner tube comes in: Cut off the part with the valve, and you'll have a long, hollow tube. (A little powder might come out.)

Create a loop to go around your arm at the shoulder and secure it to the bottom servo mount. Then, thread your doubled wire through the tube. It shouldn't go all the way through. Split the top end of the inner tube and tie it to the base of your robot (Figure Q).Now crisscross the tube across your chest to create a harness.

I used a soft bandana to wrap the shoulder loop and servo mount, which cushions them and makes the harness much more comfortable (Figure R). I also knotted it around the Arduino and its cable, to support it against strain — I plan to add a 3D-printed enclosure as well, but it will still hang down the back inside the scarf knot.

OWL SYSTEMS GO
Your robot is ready to wear. Place it on your shoulder, with the loop around your arm, then wrap the wire and tube around your torso so it's sturdily supported. Keep wrapping and tuck or tie the end of the rubber so that it doesn't get tangled with your wires. Have fun! ◐

La Petite Press

Pull traditional intaglio prints on a mini press you made yourself!

Written by Martin Schneider

Get a taste of traditional printmaking with the world's first 3D-printed etching press! With this free and open project, these unique art techniques are becoming accessible to a lot more people, and in places where they weren't possible before.

Printmaking has been around for 500 years, used for illustrating books, printing bank notes, and duplicating famous paintings. Today, printmaking is used by artists for its distinct look, but lots of people don't have access due to the high costs of a press. Etching presses for intaglio printing

MARTIN SCHNEIDER
is a 22-year-old freelance illustrator, design student, and maker from Cologne, Germany. After creating the Open Press Project, he's on a mission to make printmaking accessible to everyone.
martinschneiderart.com

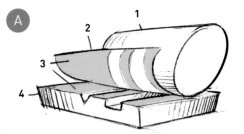
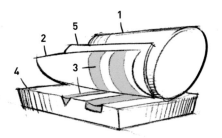

Comparison between relief printing (left) and intaglio printing (right):
1) Roller, 2) Paper, 3) Ink, 4) Plate with grooves and scratches, 5) Felt blanket.

are especially hard to find due to the high amount of pressure that's needed.

So I designed a working intaglio press that can be produced with a 3D printer: the Open Press Project (openpressproject.com). The 3D files are free to use, and the project has already grown into a small community of excited makers and printmakers all over the world.

GET IN THE GROOVES

I like to explain intaglio printing technique in comparison to relief printing (Figure A), because everybody knows how potato prints and rubber stamps work: In *relief printing*, the raised surfaces of the stamp or plate are covered with ink while the grooves remain un-inked. With *intaglio printing* it's the exact opposite: We carve scratches and grooves into a plate and fill them with ink. Then we remove the ink from the surface, so that it remains only in the grooves. The paper is then compressed into the grooves to pick up the ink.

BUILD YOUR PRINTING PRESS
1. 3D-PRINT THE PARTS

I'm using standard PLA since it doesn't warp as much as ABS, which is very important for the rollers. I use support structures for the lower roller only; the other parts should be printable without.

Keep an eye on the printing orientation (see "Perfecting Placement," page 70). Since the strength of 3D-printed parts depends on how the layers are arranged, it's important that the turquoise and orange parts are printed lying down (Figure B). The side pieces (turquoise) have an interlocking part in the middle, "printed in place," which needs to slide up and down later. You might need to break these free after printing, since they might be fused together slightly.

Print the two rollers standing up for a better surface finish; this is important for the quality of the intaglio prints later. The bed is printed with the gear teeth facing upward. The side pieces and the larger

bow-tie connectors are the exact same model; print two of each.

I suggest you use a lot of infill for the parts, since they need to withstand a lot of pressure. I'm using 25% for the side pieces and the bed, 50% for the rollers and pin, and 15% for the handle and connectors. I suggest a 0.2mm layer height for the side pieces and a maximum of 0.3mm for the other parts.

2. TRUE THE ROLLERS AND BEARINGS

Some 3D printers are not super accurate, so measure the roller diameter after printing (Figure C) and make it as round as possible with the sandpaper (Figure D). Same goes for the sliding bearings in the side pieces; if these aren't accurate, the rollers won't run smoothly later.

Martin Schneider, Sven Buchert

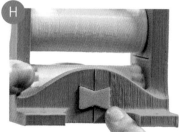

3. ADD PRESSURE SCREWS

Put an M5 screw through the top of each side piece and through its hex nut (Figures **E** and **F**) and make sure it hits the sliding bearing block in the middle. With these you'll be able to change the pressure that's applied to the printing bed.

4. ASSEMBLE THE PRESS

First make sure the rollers turn in the bearings without jamming. If they do, sand them again. You can add a dab of Vaseline to make them run really smoothly.

Insert the rollers between the two side pieces (Figure **G**) and connect everything with the little bow ties (Figure **H**).

Then push the roller pin into the lower roller (Figure **I**) and add the handle (Figure **J**). Your press is assembled.

5. MOUNT PRESS TO BOARD

I recommend that you attach your press to the edge of a wooden board before using it (Figure **K**). This makes printing much easier, since you don't need to hold the press down while you're turning the handle.

That's it! Your mini intaglio press is ready for printing.

PRINT SOME ETCHINGS

6. PREPARE THE PAPER

Before you etch a plate, first prepare the paper that you'll print on later. It's important to use special intaglio paper that can soak up the ink from the scratches on the plate.

Use tap water and a brush to wet the paper on both sides. Let it soak in a while before you use it for printing. If you're planning to make multiple prints, stack your papers and wrap them in plastic to keep them wet.

7. ENGRAVE YOUR PLATE

This might be the most difficult part: You need to come up with a subject for your plate!

After you figure out your image, grab your beverage carton (Figure **L**). It will print just like a copper plate, but it's much more accessible and easy to use.

To etch your image, use the drypoint needle to make deep scratches into the inner layer of the carton — the plastic/aluminum layer (Figure **M**). The paper backing underneath will later soak up ink and transfer it onto your print paper. The scratches are perfect if you can feel them when you move your finger over them.

The awesome thing about the beverage carton is that you can easily cut away parts of the plate entirely, or use a scalpel or X-Acto to carefully remove parts of the first layer only. That's great if you want to have large colored areas, since the ink will stay in these parts of the plate.

One more thing to keep in mind: The plate will be mirrored in your prints — everything you scratch into the plate will be flipped. That's important if you want to print text, for example.

8. INK THE PLATE

Now for the fun part: Generously cover the whole plate with ink, and fill the scratches especially (Figure **N**). Since intaglio ink is oil-based you'll have a hard time removing it from your fingers and clothes, so you might want to use gloves and an apron.

9. WIPE THE PLATE

Now use scraps of paper to remove the ink from the surface of the plate (Figure **O**). If you have an old telephone book, this is a great way to get rid of it. There will always be some ink left on the surface, but try to wipe it as clean as possible while making sure that ink stays in the scratches.

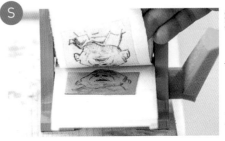

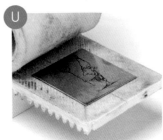

10. SET UP YOUR PRESS

Carefully place your inked plate on the printing bed face up (Figure P). Then place the wet paper on the plate and put the felt blanket on top of that (Figure Q). Your sandwich is ready to be printed.

11. PRINTING TIME!

Carefully push the sandwich in between the rollers and wait for the lower roller to get ahold of the gear teeth of the printing bed (Figure R). The teeth are angled so that there's only one orientation that'll work. You might need to remove some pressure by loosening the screws to make more space for the bed.

Now turn the handle until the bed goes through to the other side. Make sure the felt is moving (not stuck) and the upper roller is turning, too.

Now for the magic moment: Carefully lifting the felt and paper is my favorite part of the whole procedure (Figure S).

12. INSPECT AND ADJUST

Congratulations, you have successfully printed your first print on a 3D-printed printing press! If the print is too light and doesn't show clean edges, you might want to tighten the pressure screws. Also it may be the fault of your plate: Make sure the scratches are deep enough.

Experiment with different amounts of pressure and the wetness of the paper to get your best prints. In Figure T you can see the differences between two prints. While both are quite good, the lines of the left one are more blurry while they're super sharp on the right one. For the left one, the surface of the plate had more ink left on it, that's why there is more color in general. But this comes down to preference — you just have to figure out what you like.

13. DRY YOUR PRINTS

Place each of your prints in between sheets of regular copy paper to make a stack, then place some heavy books on top. Let this sit for a few days, replacing the copy paper with some fresh paper after a day or so. After that, you're ready to sign your prints!

KEEP IT ROLLING

Now that you're a printmaker, there's much more to explore. For your intaglio plates, you can use all kinds of materials including copper, zinc, aluminum, steel, plastics, CDs, linoleum, wood, and foam board. I'm using copper most of the time (Figure U), though for beginners it's less expensive to practice on juice boxes or CDs.

Your press can also be used for relief printing — linocuts, woodcuts, and 3D-printed plates like those used by Christopher Sweeney (Figure V) work great, as does mixed-media collagraphy, which can be printed intaglio or relief.

I would love to welcome more makers to our community on Instagram: instagram.com/openpressproject. Please share your presses and prints, it would make my day!

I'd also love to see your redesigns, adjustments, and improvements to the intaglio press itself. After 300 years with little innovation, it's time to take it to the next level. All the STL files are licensed as Creative Commons Attribution-NonCommercial, so you're free to copy, share, and edit the press for non-commercial purposes.

Questions or feedback? Email me at martinschneiderart@gmail.com. I would love to hear from you! ●

[+] Get more photos, and tips on sourcing materials, on the project page at makezine.com/go/3dp-printing-press.

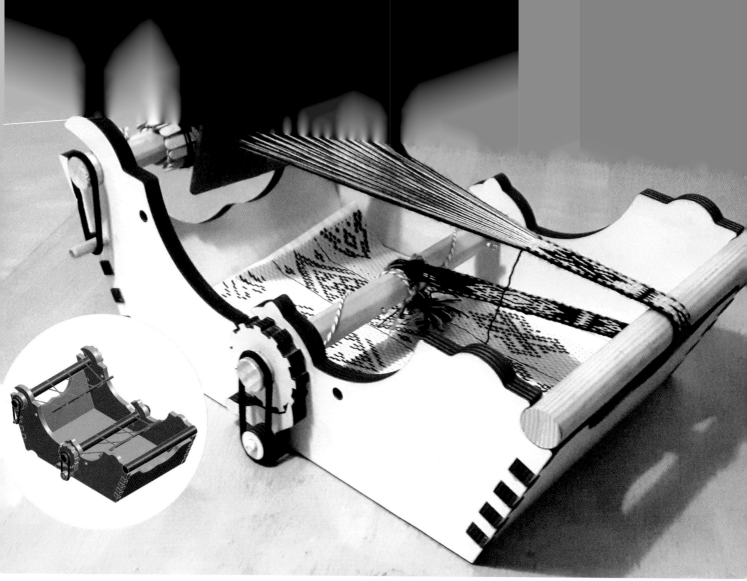

Laser-Cut Box Loom

Written by
Matthijs Witsenburg

A portable, easy-to-build machine for band or card weaving

My wife showed me some pictures of old Norwegian box looms and asked me to make something similar. The reason being that a back-strap loom, worn on the body, can be impractical with small children running around. I'm not much of a woodworker, so I decided to design something that could be laser cut and then assembled with a minimum of work.

This box loom is designed for use with a rigid heddle or for card weaving. It can be used on a table or on your lap, and the work can be dropped at a moment's notice when something else (kids, pets, etc.) requires attention right *now*.

The cloth and warp beams are held in slots rather than holes. This makes construction easier, and, as a bonus, you can remove and set aside an unfinished

project if something more interesting comes along. Simply release the tension, carefully take out the beams, roll up your project, and secure it with rubber bands.

If you build more than one loom, the recesses on the lower edge allow for easy stacking.

Once you have the laser-cut parts, the rest of the loom can be made with the most basic of woodworking tools. All measurements are in millimeters, but none are so critical that they can't just be eyeballed.

1. LASER-CUT THE PIECES

On the cutting plan, the black lines must be cut, the red lines etched. Import the file in millimeters, not inches. The clearance for the finger joints is 0mm.

The kerf from the laser should be sufficient for these joints. If you were to use a CNC router instead of a laser cutter, you may want to widen the joints or play with the thickness of your bit.

All parts are nested in a 500mm×500mm area. If you use solid wood rather than ply, keep in mind the wood grain and move the pieces around as needed.

2. SHAPE THE PANELS

The bottom, front, and back panels will need some shaping. All edges that need shaping have an etched line, indicating where the beveled edge starts. All sides with an etched line will end up on the outside of the loom (Figure Ⓐ).

Clamp each panel to a work surface and remove material outside this line. I prefer to

Matthijs Witsenburg

TIME REQUIRED:
3–4 Hours

DIFFICULTY:
Easy

COST:
$15–$60

MATERIALS

» **Laser-cut parts, in 12mm plywood**
Download the plans and DXF files for cutting at hackster.io/matthijs-witsenburg/laser-cut-plywood-bandweaving-loom. I used poplar; most types of solid or plywood will work. You can substitute 10mm or ½" plywood; you'll just have to adjust the beveled edges.
» **Wood dowels, 22mm diameter:** 24cm long (1) and 29cm long (2 or 4)
» **Wood dowels, 8mm dia., 40mm long (4)**
» **Strings, 75cm long (2)**
» **Hair ties (4)**
» **Wood glue (PVA), white or yellow**
» **Teflon tape, thread, or cling film (optional)**
FOR WEAVING:
» **Yarn, shuttle, and a rigid heddle or weaving cards**
» **Pencils (2)** or similar thin rods about 8mm dia. and 21cm long, for apron rods

TOOLS

» **Laser cutter or CNC router**
» **Wood saw** to cut dowels
» **Clamps (2)** at least 24cm opening
» **Drill with 3mm drill bit**
» **Plane or wood rasp**
» **Sandpaper or sander**
» **Plastic bag**
» **File (optional)**

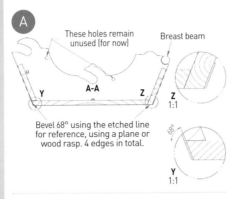

These holes remain unused (for now)
Breast beam
Y   A-A   Z
Z
1:1
Bevel 68° using the etched line for reference, using a plane or wood rasp. 4 edges in total.
Y
1:1

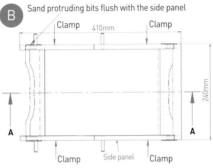

Sand protruding bits flush with the side panel
Clamp   410mm   Clamp
240mm
A   A
Clamp   Side panel   Clamp

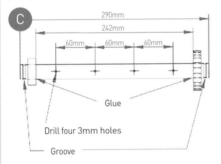

290mm
242mm
60mm 60mm 60mm
Glue
Drill four 3mm holes
Groove

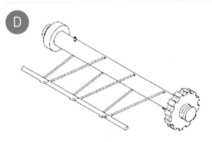

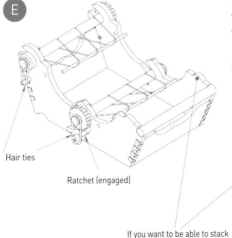

Hair ties

Ratchet (engaged)

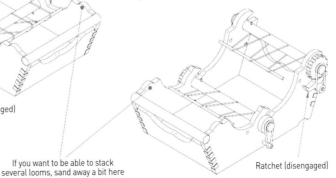

If you want to be able to stack several looms, sand away a bit here
Ratchet (disengaged)

use a wood plane for this. A rasp or coarse sandpaper will work as well.

3. GLUE THE LOOM BODY

First, do a dry fit to see how the panels go together and where the glue actually needs to be. Spread a thin layer of carpenter's glue on one side of all the panel joints, assemble, and clamp as indicated in Figure **B**, using just enough pressure to keep the pieces in place. Set to dry on a plastic bag.

Take a step back and admire your work. If it appears to be crooked or put together wrong, adjust or take it apart and redo. If there's any glue squeeze-out, wait until it becomes rubbery and scratch it off.

Put a drop of glue in the holes for the 8mm pins and push in the pins in a twisting motion, holding a finger on the back of the hole to prevent pushing out the glue. Spread some glue in the recesses for the breast beam and place the 24cm dowel. No clamp is needed here.

Once the glue has set, use sandpaper to round off the edges to your liking. I prefer to sand down the protruding bits of the finger joints, but that's mainly cosmetic.

4. MAKE WARP AND CLOTH BEAMS

Make a pair of these (Figure **C**), or more if you want interchangeable sets. The cutting plan has enough parts for 2 pair.

To drill the holes (marked in red), clamp the 29cm dowels to a sacrificial piece of wood. Don't worry if the holes aren't very straight. If you wish, you can also file a small groove around the ends of the beams for the hair ties to sit in.

Test-fit before gluing the beams; they should easily turn in the slots with a small amount of play. But don't leave them in the loom to dry, as they might get stuck there.

For the apron strings, loop a 75cm string through the holes as shown in Figure **D**. Make a knot only in the ends. This way the length of the loops around the apron rods can be equalized afterward.

5. FINAL ASSEMBLY

Place the warp beam and cloth beam in the slots (Figure **E**). Place the ratchets on their pins and secure them with the small rings. Don't glue these; if the small rings are too loose, you can thicken the pins by wrapping tape, thread, or cling film around them. The ratchets should move freely around the pins. If they don't work to your satisfaction, put some extra 8mm dowels in the unused holes and use these for the ratchets.

The hair ties keep the beams in place while warping the loom. Remove them after warping, or leave them, whatever works best. You can see different warping setups at makezine.com/go/laser-cut-loom.

I also made a heddle from 3mm Delrin (POM) plastic. Laser cutting POM leaves very smooth edges that won't fray your warp.

Now all that remains is to pick a project and get weaving. Have fun! ✪

3DP
Written by Tasker Smith
Leather Press

Print your own custom tools for traditional leather forming

TASKER SMITH is a technical instructor at MIT who mentors students in the practical use of digital fabrication technologies and the process of iterative prototype development.

TIME REQUIRED:
4–5 Hours

DIFFICULTY:
Intermediate

COST:
$20–$40

MATERIALS
» Vegetable-tanned cowhide, 4oz
» Leather dye, water-based
» Leather conditioner

TOOLS
» 3D printer
» Clamps
» X-Acto knife with #13 blade
» Cotton swab

Leather forming is a craft that dates back to 3000 BCE when ancient civilizations began using animal hides for bags, garments, boots, and sandals. Five millennia later, we still use animal hides for all these purposes, but the technology available to craft leather has evolved with our culture. For this project I created 3D-printed tools to make a custom leather press, complete with an imprint of Makey.

1. DESIGN
I started by designing in SolidWorks the finished component I wanted to create out of leather. I designed a solid body with a square profile and edge fillets, followed by an extruded detent in the shape of Makey. After the exterior surface detail was to my liking, I "shelled" the model to the thickness of the leather I intended to use. This resulted in a solid model of the leather component I wanted to create.

Once this model was complete, it was relatively easy to use the Mold Tool in Solid-Works to craft a two-part mold (Figure Ⓐ).

2. 3D PRINTING
Any 3D printer will work here, but it's important to note that surface imperfections and build lines translate directly into the surface of the leather. Set your printer software to the finest layer resolution, and orient your components so as to minimize build lines on finish surfaces (Figure Ⓑ). Also, spend some time post-processing the components to refine the details you want to see in the leather and eliminate any raised imperfections you don't (Figure Ⓒ).

3. CHOOSE LEATHER
Vegetable-tanned cowhide is the industry preference for shaping leather: It tends to hold shape better than hides of other animals or those tanned with a chromium solution. It's generally a pale brown color, though a variety of dyes are available for tinting. More on this later.

Leather is processed to a uniform thickness, and is named according to the weight of a square foot of material. For example: 1 square foot of "4 ounce"

Tasker Smith

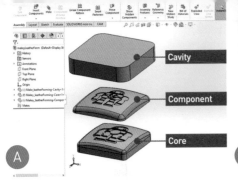

A

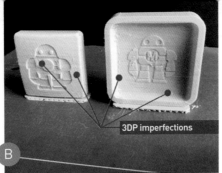

B — 3DP imperfections

C

D

E

F

G

H

leather weighs 4oz. I cared more about thickness than weight, and had to consult a chart to source the 1/16" thickness of material that I wanted for this project.

4. WET THE LEATHER

The goal is to soften the leather with warm water, form it into the desired shape, and allow it to dry and harden such that it holds the shape of the form tool. A few factors are important to consider:

» If the leather is too wet or too dry, the shape won't hold. Make sure moisture penetrates the full thickness of the hide, but you don't want it soaking wet.

» Not too hot! Boiling leather causes it to shrink and harden considerably. This may be desirable for some projects, but in this case it was important to keep the temperature below 160°F.

5. PRESS THE LEATHER

Once the leather is moistened, clamp it tightly in the form for a few hours to dry (Figure D). I found that 2 hours was

sufficient to hold the shape (Figure E), though you may want to go longer if fine detail is lost.

6. TRIM

I designed the mold so that I had an edge to trim against, and left the clamps in place during trimming to prevent any shifting of the material. Using a #13 X-Acto blade, I was able to trim through the leather in a single pass (Figure F). Depending on the thickness and hardness of your leather, you may need to score it with a pass or two before fully trimming off the excess.

7. FINISH

A variety of colors are available to dye leather. I used a water-based dye, testing my application process with scrap material before applying tint to the finished part (Figure G). The folks at Tandy Leather recommended applying dye with a cotton swab in a circular motion; I found this worked well to eliminate evidence of brush strokes. To get the desired

color saturation (Figure H), I found I needed to dilute the dye with water.

As a final step, I used a leather conditioner. This helps to adjust the surface sheen and seal in the color of the dye to eliminate premature wear.

THE LUXURY OF LEATHER

The results can be quite beautiful. 3D-printed objects are often criticized for feeling cheap or flimsy, but using them as tools can lead to creative solutions with a variety of materials. By marrying the versatility of digital design and fabrication with luxurious materials like leather, you can supercharge your process and generate customized artifacts worthy of handing down from generation to generation. ◆

[+] See more photos and share your 3DP leatherwork at makezine.com/2018/07/09/use-3d-printing-shape-leather

Functional Furniture

Written by Star Simpson

Updating a classic kids' table and chair design for cutting with the Shaper Origin handheld CNC router

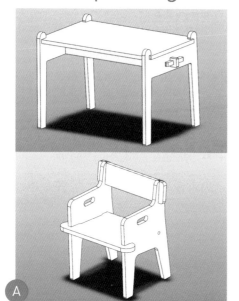

STAR SIMPSON is an engineer, glider pilot, and lifelong maker.

A few years ago I accepted the honor of becoming godmother to a charming child named Tara, who has grown into the cutest toddler I've ever met. About this same time, I found myself fascinated by furniture design, reading endlessly and looking at pictures of chairs and tables and trying to pick out all the thoughtful nuances their creators had infused into their designs. I became particularly interested in Danish craftsman Hans Wegner (1914-2007), whose chairs stood out to me, each one seeming to bear a spookily intuitive take on human comfort while also embodying creative, organic, and beautiful forms that furthermore showed off the finest qualities of the wood. In order to more fully appreciate his genius, I resolved to find time to build at least one of his designs.

During my research I found that Wegner had created a child's table and chair set, using principles of what we today call *knock-down* or *flat-pack* design. This set, which is still manufactured in solid beech

hardwood, was intended for his friend's young son Peter. The principle behind Peter's Table and Peter's Chair is that they can serve several functions: as tasteful and functional furniture for the child, but also as toys — things the child can take apart and reassemble without tools, giving a sense of ownership and a spatial understanding of their construction — while also packing away easily for the sake of the parents.

I knew immediately what my target project was to be: Tara's Table & Chair. You can try my process to re-create or update your own flat-pack furniture favorites.

1. CREATE 2D FILES FOR CUTTING
Luck was with me when I found 3D STL files and 2D dimension drawings for the furniture provided online by the manufacturer (carlhansen.com/en/collection/childrens-furniture/ch410 and /ch411). These aren't directly useful for fabrication, but provided helpful guidelines for creating my own plywood versions for CNC cutting — a

TIME REQUIRED:
1–2 Hours

DIFFICULTY:
Easy

COST:
$100–$150

MATERIALS

» **Plywood, ¾" thick, 4'×8' sheet, appearance grade** I used "ACX"-grade maple-skinned plywood. The A (appearance) side is sanded, the C side is rough (avoid big knots or voids), and X means it's durable (exterior rated).
» **Wood dowels, ⅜" diameter, 2" long (2)**
» **MDF (medium-density fiberboard), 4'×8' sheet (optional)** for a spoilboard, if you're using the Shaper Origin router
» **Tung oil**
» **Furniture wax**
» **Wood glue**

TOOLS

» **Computer** Download the cutting files from the project page, makezine.com/go/taras-table.
» **CNC router (optional)** I used the Shaper Origin handheld CNC router, but you can also use a flatbed CNC or a drill and handheld saw.
» **Wood chisel**
» **Handsaw**
» **Center punch**
» **Drill and bits**
» **Sander and sandpapers**

We set the plywood up in the garage on sawhorses, clamped the sheets in place, and set to work applying Shaper's special visual fiducial tape, which resembles a non-repeating game of dominos, in order to use the Origin for our cutting (Figure B).

After a few hours of cutting, including breaks for lunch, the correct two-dimensional forms were extracted from the plywood sheet and test-fit together (Figure C).

3. CLEAN UP CORNERS AND EDGES

A few of the inner corners were not possible to cut perfectly with a router bit, so I used a chisel to make these exactly 90° (Figure D).

Next, I smoothed all the sharp edges left by the router to make them suitable for tiny hands, using a handheld sanding block to break down the inner edges, and a power belt sander and disc sander to do the rest. I first took every 90° edge down to a 45° beveled edge (Figure E), then used sandpaper to round them.

4. GLUE THE SEAT PINS

In the original design, the hardwood seat is carved with a pin on each side, to fit into holes in the sides of the chair. I used dowels instead. I turned the plywood seat on its side and measured the correct location for each dowel, used a punch to mark the center, drilled a ¼" pilot hole, and then drilled a ¹⁵⁄₃₂" hole about ¾" deep. Into these holes, I poured a generous amount of wood glue, then pressed in a ⅜" dowel, making sure about 1" would protrude, to extend the full width of the chair's side but not much more.

5. KID-FRIENDLY FINISH

To make the furniture more or less crayon- and juice-proof and to bring out the light golden glow of the maple, I purchased a pint of tung oil and used most of it. I daubed it on with a cheesecloth, applying no more than the surface would seem to absorb and then letting it dry for a few hours before applying the next coat, achieving 4 separate coats to each piece of wood (Figure F).

I then rubbed furniture wax in over every oiled surface, to seal in the finish and create a shiny, glassy depth.

PRACTICAL AND PLAYFUL

After giving this finish a few days to cure, I delivered to Tara her new table and chair, which we enjoyed putting together and then immediately putting to use! ●

Keith Cormier, Adem Rudin, Star Simpson

process that otherwise would have required careful attempts to measure dimensions from photos online. I called on a mechanical engineer pal, Adem Rudin, who very generously helped me to convert the 3D files to 2D CAD files — first in the DXF format and then as SVGs (Figure A). You can download these from the project page.

You can also find dimension files on many furniture manufacturer's sites, and find flat-pack designs on SketchUp 3D Warehouse, Thingiverse, and other sharing sites.

2. CUT PIECES WITH CNC

To machine the wood using these files, you could use a 2D CNC router like a ShopBot, or even a drill and a handsaw. I was able to borrow a Shaper Origin from a nearby hackerspace. It's a handheld router with a camera and CNC "autocorrect" feature, which smooths out the inevitable jerks and jitters and makes it possible to cut wood manually with CNC accuracy.

Armed with my design files and CNC tool, I set out for the hardware store to select some ACX-grade maple-skinned plywood and a sheet of MDF to use as a spoilboard.

The Big Picture

Written by Matt Bell

Use PolyProjector to cut and build enormous 3D models out of cardboard

For Maker Faire Kansas City this year I wanted to make a drone racetrack for micro FPV quads. Normally these courses consist of lots of LED-lit circles for the quads to fly through. I wanted something much more sculptural and ambitious: giant spider robots mining a floating asteroid field (Figure A). To keep costs down, and because it's generally an awesome building material, I decided to use cardboard. My new laser cutter had just arrived and I figured I'd use it to do all the cutting.

But a problem arose. I wanted giant asteroids 4 to 6 feet in diameter, and the pieces were just too big for my little laser cutter. I needed another way to make them.

The approach I decided on was to create 3D models of my asteroids and then use a digital projector to cast the individual polygons onto big sheets of cardboard so I could use ordinary tools to cut them out.

Possessing the superpower that is knowledge of software development, I decided to make my own little program to make this process possible. It's called PolyProjector. Here's how you can use it to make giant cardboard forms of your own.

1. FINDING BIG CARDBOARD

You can save up Amazon boxes or ask at the grocery store, but the best source I know is a neighborhood recycling center. The ones I've been to have giant walk-in shipping containers filled with clean, stacked cardboard (Figure B). Not just little stuff either — they've got the big boxes that refrigerators and couches come in.

You'll probably want to get a consistent thickness of cardboard. Turns out there's a huge variety of the stuff: lightweight, heavy duty, single ply, double ply, even stuff that's four layers thick. You can mix and match; you'll just have to adjust on the fly during assembly of your model.

2. PREPARING THE 3D FILE

The program I wrote accepts OBJ files (sorry, no STL) and expects the units to be in meters. If you need to convert files from another format or change the scale, check out Blender. It's free, can import many file types and export OBJs, and works on just about any computer.

3. PROJECTOR SETUP

Set up your projector so that it points down at a horizontal work surface and covers

TIME REQUIRED:
1–2 Hours

DIFFICULTY:
Easy

COST:
$10–$20

MATERIALS
» **Corrugated cardboard** the bigger the better
» **Tape** I used 2" Shurtape FP-96 General Purpose Kraft Packaging Tape. It worked well.

TOOLS
» **Digital projector**
» **Computer with PolyProjector software** free download at github.com/greengiant83/PolyProjector
» **Long straightedge** like a yardstick
» **Sharpie marker**
» **Cutting tools for cardboard.** You can use shears, a box cutter or utility knife, or my choice, a table saw.
» **Scissors**
» **Pencil**
» **Rubbing alcohol** to clean tape gunk off your scissors

MATT BELL runs an innovation lab for a global advertising firm out of Kansas City. He loves long walks on the beach and elegantly engineered bits of code.

enough area to project the biggest of your pieces. You'll have to do a bit of MacGyvering to get it in the right spot. My solution was to bolt the projector to a piece of plywood, then hang the whole thing from pegboard in my shop (Figure **C**). Whatever technique you use, double-check that your mounting is secure (so you don't have any tragic projector-maiming accidents) and isolated from your work surface (so little bumps won't translate into time-killing wobbles of your projector image).

Next, calibrate your projector so that when the computer tries to project a 4" square it actually comes out as a 4" square. When you launch the PolyProjector program it will project a large cyan rectangle to show the work area, and a red rectangle labeled Notebook Paper. Put a standard 8"×10½" sheet of notebook paper in the red rectangle and move the corners around until the projected rectangle lines up with the physical one. This step is a little finicky, but you only have to do it once. The program will save your settings for next time.

4. MARKING AND CUTTING PIECES
Now for the fun stuff — cutting some actual pieces. Press L to load your 3D model. This brings up a view with the 3D model on the left, and first of the polygons to cut in the center. You can toggle the 3D view on and off using the Tab key. Click and drag the model to rotate it, or use the scroll wheel. All the faces are numbered; you can cycle through them using the < and > keys (comma and period) to bring a specific face into view (Figure **D**).

Once you've selected a face, use your scroll wheel to rotate the polygon, and click and drag to move it into a good position on your sheet of cardboard. You have your first shape glowing on the cardboard (Figure **E**)!

Take a nice long straightedge and trace each edge using a Sharpie. Take care not to cast shadows on an edge and mark the wrong spot. Apply a little pressure to the straightedge to make sure the cardboard lies flat, so that the projection is accurate. Lastly, use a pencil to mark the face number. This will be crucial when it comes time to assemble everything.

Once the cardboard is fully marked, cut it out. You can use a variety of knives, but I use a table saw with the fence removed. It's quick and it yields nice straight lines.

Press the > key to move on to the next face. Repeat until you've cut all your shapes.

5. ASSEMBLING EVERYTHING
Go back to the 3D view of the model. Using this view as a map, you can start taping all your pieces together. Aren't you glad you marked the face numbers?

Use lengths of tape that run the full length of the edge. Structural rigidity increased noticeably when I taped both sides of the cardboard. If you cut the end of the tape with two cuts so that it's pointy in the center you can minimize overlapping and make things look much cleaner.

POOR MAN'S LASER CUTTER
So there it is — a tool for projecting the individual faces of a 3D model onto sheets of cardboard to create physical models, jumbo size (Figure **F**). It's kind of like a poor man's laser cutter.

I spent 3–4 weeks working with cardboard for my exhibit and it left me with a fine appreciation for the material. It is practically free, can be easily and quickly shaped using a handful of tools, and as long as you don't let it get wet, it'll last quite some time. ◎

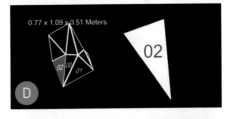

[+] See more photos and tips at makezine.com/go/giant-cardboard-projector.

1+2+3 Easy Soda Can Robot

Written by
Merve Güngör

In this project we'll make a walking robot using a Coca-Cola can and a little trick for gluing two servos together. It's the easiest way I know to create a bipedal gait. Let's start!

1. GLUE THE MOTORS

Cut the yellow mounting tabs off both motors, then glue the flat sides of the motor shafts together as shown in Figure A. Make sure the glue doesn't touch the yellow parts! It's this off-center shaft connection that creates the walking motion.

Wait for the glue to cure, then glue the batteries to the bottom of the motors — they'll be the robot's feet (Figure B).

2. WIRE THE CIRCUIT

Solder the circuit as shown in Figures C and D, then test the robot by closing and opening the switch. If the robot isn't running, it means you made a mistake in the circuit. Check connections again. Once it's working, you can protect your connections with more glue.

3. DECORATE YOUR ROBOT

Now the most fun part: Design your robot as you like!

Cut the bottom off the Coke can. Cut a small rectangular notch in the can and glue the switch into it. Then glue the can on top of the motors (Figure E).

To make the robot's head, stick the can bottom on the top of the robot. I used googly eyes (Figure F), a drinking straw for arms, and pipe cleaners for hair. But you can decorate your robot however you want!

GET MOVING!

Your robot is ready. Let's see how it works and walks! ◗

TIME REQUIRED:
1–2 Hours

DIFFICULTY:
Easy

COST:
$5–$10

MATERIALS
» **Hobby gearmotors, "TT" type, 6V, 1:48 ratio, 90° shaft (2)** such as Adafruit #3777, adafruit.com
» **9V batteries (2)**
» **9V battery clips (2)**
» **Hookup wire**
» **Soda can, 12oz**
» **Switch**
» **Materials for decoration**

TOOLS
» **Hobby knife**
» **Soldering iron & solder**
» **Hot glue gun**

MERVE GÜNGÖR
lives in Istanbul, Turkey. Her motto is "Make what you love, share what you make."

[+] You can watch Merve's video and share your builds on the project page at makershare.com/projects/lets-make-walking-robot-coca-cola-tin. Questions? Contact her at mrvgngr.0@gmail.com.

Give the Gift of Inspiration

Gift one, Get one

Make:

Go to makezine.com/subscribe
to get started

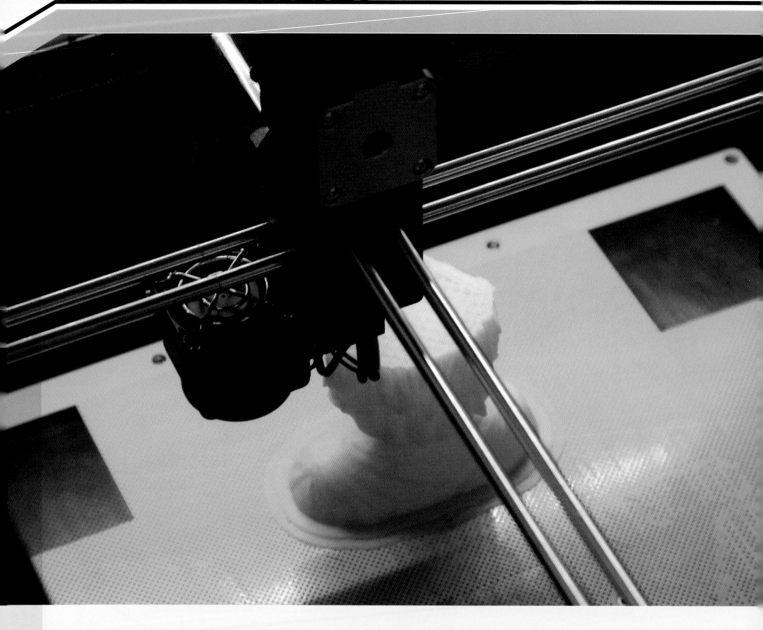

Perfecting **Placement**

How to position your 3D model on the digital build plate for optimal results

Written by Lydia Sloan Cline

LYDIA SLOAN CLINE teaches 3D printing, Autodesk programs, SketchUp, and board drafting at Johnson County Community College in Overland Park, Kansas. She creates video courses, holds SketchUp and 3D printing workshops, judges at competitive technology events, and loves all things technology. Find tutorials of her books' subjects at **youtube.com/profdrafting**.

This tutorial is excerpted from Fusion 360 for Makers, available at makershed.com and other retailers.

FUSION 360 IS A ROBUST 3D DESIGN PROGRAM THAT ENABLES YOU TO DO *COMPUTER-AIDED DESIGN* (CAD), *computer-aided manufacturing* (CAM), and *computer-aided engineering* (CAE) tasks. It can animate (make videos), render (add lifelike colors and textures), and display the model as a set of scaled, 2D drawings. It also lets you arrange the model's parts into an assembly. Whether you're new to modeling, a hobbyist, or an experienced engineer, you'll find this software useful.

You can export any Fusion 360 design as an STL for 3D printing. However, it won't necessarily print, or print well, even if it looks good on the screen. For best results, all designs should be *optimized* for printing — here we will focus on orienting the file on the digital build plate.

> **NOTE:** Some suggestions may conflict with others, depending on the model. You may need to ignore one suggestion to implement another.

ORIENT TO AVOID OR MINIMIZE SUPPORTS

Supports are structures that hold up *overhangs* (parts with empty space below). Slicing software generates supports, but you can finesse by adding, deleting, or moving them to other locations. After the print is finished, remove supports by snapping them off, cutting them off with a craft knife, or dissolving them, if they were made with dissolvable filament.

Supports are often difficult to remove, or fail during the printing process, so try to orient the model to avoid or minimize them. Features that are at a 45° or greater angle don't need supports, because each layer is built onto the one underneath it (Figure **A**). Simply rotating the model can also cut down or eliminate the number of supports needed (Figure **B**). You may be able to incorporate supports in the design, or slice the design in half with the Split Body tool to allow each part to lie flat on the build plate, and then glue the printed parts together.

ORIENT FOR STRENGTH

Print items that need to be strong laterally on the z layer, not vertically across multiple layers. For example, if you print the screwdriver handle in Figure B vertically, it

will easily snap at each layer. But printed horizontally, the layers span the entire length, making the handle stronger.

ORIENT TO MINIMIZE WARPING

The build plate is most level at the center, so parts printed there will warp less than parts printed closer to the edges. Heated build plates are also hotter at the center. Cooler print locations contribute to warping.

ORIENT TO AVOID STAIR-STEPPING

Print sloped and curved surfaces horizontally. Printing them vertically results in a *stair-stepping* effect, especially when the surface has a shallow angle (Figure **C**).

ORIENT TO PRESERVE DETAIL

Thin layers print better vertically than horizontally. So position thin, detailed models like *lithophanes* (3D printed photos) vertically, as shown in Figure **D**.

On a printer with a moving bed, layers close to the build plate print best because they sway less than layers higher up. This means that highly detailed features print better if they're oriented closer to the build plate. For example, the mane on the lion in Figure **E** will print best if it's close to the build plate as shown.

Supports often leave scars when removed, so orient a file that has detail you don't want ruined in a manner that avoids supports resting on them. Figure **F** shows how tilting the file 45° eliminates most of the supports that would otherwise mar the detail.

ORIENT FOR SMOOTHNESS

A glass build plate gives the bottoms of prints a very smooth finish. Most of the time, this nice finish is wasted on the bottom of the print's base. If you have a file with a long outside face, such as a phone cover, you might want to orient it so it rests directly on the glass.

GOING FURTHER

To learn more about the software and other best practices for successfully printing your digital model, pick up a copy of *Fusion 360 for Makers* at makershed.com or other retailers. ⊘

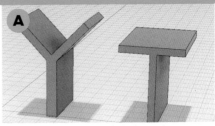

The 45° features on the left don't need supports; the perpendicular overhangs on the right do.

The screwdriver handle oriented straight up doesn't need any support; the screwdriver handle oriented horizontally requires a lot of support.

Stair-stepping on a vertically printed model.

Position a lithophane vertically for best detail printing.

Position detailed features closest to the build plate for best detail printing.

Orient the file to avoid putting supports on details.

Micro Milling

Written by Winston Moy

Move up to the next level by going small on your CNC machine

WINSTON MOY is a self-taught hobby machinist and shameless digital fabrication evangelist who shares his journey of building, crashing, and learning about CNCs on his channel youtube.com/winstonmoy.

FURNITURE, WALL HANGINGS, AND CUSTOM SIGNAGE ARE FAMILIAR CNC project categories, but downsizing your ambitions can also open up a world of possibilities. That said, machining miniature parts is not without its own challenges. Small parts are tricky to secure on a CNC, so you may need to get creative with your workholding and look beyond using basic clamps. The importance of using the right materials and tools is also elevated, as your margins for error are smaller.

WORKHOLDING

Holding your material securely is a prerequisite for success in CNC, and tiny projects can be difficult to get a grip on.

A **LOW-PROFILE VISE** is indispensable where you may not have a lot of z-clearance. They keep the top face of your part exposed (Figure **A**). However, if your part is oddly shaped, or you need complete access to the sides, then consider a different method.

ADHESIVES are great for holding thin stock with a lot of surface area. Double-sided tape is often used to secure PCB blanks, plastic, or plywood (Figure **B**).

Fixturing wax or hot glue can be a good way to secure small pieces to your table. Melt it to the surface, press your stock into the molten puddle, and maintain pressure while it cools (Figure **C**). A nice benefit with these methods is that there's no sticky residue to gum up your end mill once you cut through your part since wax and hot glue aren't tacky when cooled.

When machining parts with curved or irregular surfaces, custom fixturing solutions called **SOFT JAWS,** which are clamp faces that have the negative profile of your part machined into them, can be useful (Figure **D**). You could also cut a shallow pocket in the spoilboard to hold your parts, like in Figure **E**.

MATERIALS

Small projects are a great opportunity to start working with metals. The most common non-ferrous metal to machine is **ALUMINUM,** but not all aluminum is made equal. Different alloys and tempers can drastically affect how well it machines. Pure aluminum is relatively soft and almost acts like clay at the microscopic level — and it clogs end mills easily.

7075 aluminum, which is alloyed primarily with zinc for strength, is a harder alloy that shears much more cleanly than pure aluminum; it "forms a chip" as they say in the machining world. A cheaper alternative that's slightly softer but still very CNC-friendly is 6061 aluminum. 6061 is alloyed primarily with magnesium and silicon, and is a common general-purpose grade aluminum. Using the wrong alloy can cause you endless headaches, so do some research before buying. McMaster-Carr's website is a good reference.

In **PLASTICS,** it's important to have a good ratio of feed rate to spindle RPM. Wood will char if your cutter loiters too long in a single spot, but plastic will melt, clog your end mill, and ruin your project. Heed the recommendation of your CNC manufacturer, or better yet, the cutting parameters of your tool vendor.

END MILLS

Speaking of tools, different **END MILLS** have their own pros and cons but the one truth that holds in any situation is that you should use the shortest tool you can get away with. The more your tool sticks out of a collet, the greater your risk of vibration or "chatter."

Furthermore, the longer your flutes are, the weaker your end mill is. In a small machine, if you get your cutting parameters wrong, you'll likely stall your CNC before you break a tool, but in miniature scale

Brass branding iron head made in a low-profile vise.

Low-profile vise with a hard stop allows for easy, repeatable locating of your parts.

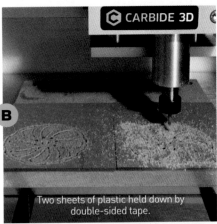

Two sheets of plastic held down by double-sided tape.

An aluminum part secured to the spoilboard by wax. No tabs means no additional cleanup is necessary.

Soft jaws can be machined to securely hold any shape, including round ones.

These drink coasters were inspired by the shape of SpaceX's drone ship barges.

machining your end mill will be the weakest link.

If you need exceptionally fine details, an **ENGRAVING BIT** can be a good alternative to using tiny, fragile end mills to define sharp corners or other tight geometry. Bonus: the engraving bit can also chamfer edges, saving you from having to file or sand off burrs.

FINISHING

Finish passes are essential for accuracy and surface finish. A duplicated cut will clean up excess material that your cutter might have missed on the first pass due to vibrations or other physical phenomena. A quick way to implement finish passes in any CAM software, basic or advanced, is by simply duplicating an existing toolpath.

A LITTLE INSPIRATION

I hope these tips help you understand some of the variables at play when tackling smaller projects, and encourage you to try your hand at making more detailed or precise parts on a CNC. ✪

G1 Z0.2

M104 S215

G90

G1 X100.0 E20.0

Kelly Egan

Start and Stop
G-Codes

*Make your printer do your exact bidding
with these commands* Written by Ryan Priore

RYAN PRIORE is an entrepreneur in the photonics industry and the co-founder of 3DPPGH, a monthly meetup of 3D printing enthusiasts in Pittsburgh, Pennsylvania.

3D PRINTERS COMMUNICATE VIA A LANGUAGE CALLED G-CODE, like all Computer Numerical Controlled (CNC) machines. This versatile language provides a set of human-readable commands for controlling each action that a 3D printer performs. Slicing software may be thought of as an interpreter for translating 3D models into a series of G-code commands for producing a solid part.

We tend to spend a lot of time focusing on key parameters affecting the final print quality (layer height, infill, perimeters, etc.), but slicing software adds two additional sections or **scripts** of G-code to the sliced files: **start.gcode** and **end.gcode**. These two highly customizable scripts have a dramatic effect on your final print.

Let's examine two typical G-code lines to better understand how a command is constructed. Commands beginning with **G** control movements and offset definitions, while commands beginning with **M** control miscellaneous actions.

G-code command	Description
G1 X25.0 Y15.5 E17.4	Linear move command (G1): Simultaneously move the tool head 25.0 units on the X axis and 15.5 units on the Y axis while moving the Extruder by 17.4 units. Units may be inches or millimeters.
M107 ; Turn off fan	Turn off the tool head cooling fan. Note that anything entered on the line after the semicolon is a comment and is ignored by the 3D printer.

START.GCODE

The purpose of the *start.gcode* script is to prepare the 3D printer for producing the desired object. At a minimum, the extruder and heated bed (if applicable) need to be set to proper temperatures and the tool head needs to be homed. To increase your probability of a successful print, you should also perform additional actions like leveling (or tramming) the bed, priming the hotend, and even updating the LCD to inform the user that the print is underway. The script below walks through a compilation of typical *start.gcode* lines that many current slicers employ:

```
G21                    ; Set all units to millimeters
M107                   ; Turn off the part cooling fan
M104 S215              ; Set extruder to 215°C [and immediately
move on]
M140 S60               ; Set bed to 60°C [and immediately move on]
M190 S60               ; Set bed to 60°C [and wait for 60°C]
M109 S215              ; Set extruder to 215°C [and wait for 215°C]
G28                    ; Move toolhead to origin (or home X, Y, Z)
                       ; Prusa uses G28 W to perform homing command
G29                    ; Auto-Level the printer bed using a
measurement probe
                       ; Prusa uses G80 to accomplish a mesh bed
leveling
G92 E0.0               ; Reset the extruder position to 0mm
G1 Z0.2                ; Move hotend nozzle to Z position of 0.2mm
G1 X100.0 E20.0        ; Prime the hotend (Move to X=100mm &
Extruder=20mm)
G92 E0.0               ; Reset the extruder position to 0mm
G90                    ; Set to absolute positioning as opposed to
relative
M83                    ; Set the extruder to relative positioning
M300 S300 P1000        ; Play a 300Hz beep sound for 1000
milliseconds
M117 Printing...;      Update the LCD screen with "Printing..."
```

END.GCODE

The purpose of the *end.gcode* script is to ensure that all printing functions have been halted, and perform any final cleanup tasks. From a safety standpoint, the most important job is to turn off power to the heating elements and motors. The script below walks through a compilation of typical *end.gcode* lines:

```
M107                   ; Turn off the part cooling fan
G28 X0                 ; Home X axis and remove hotend from object
```

```
M104 S0                ; Turn off the extruder [and immediately
move on]
M140 S0                ; Turn off the bed [and immediately move on]
M84                    ; Turn off the stepper motors
M300 S300 P1000        ; Play a 300Hz beep sound for 1000
millisecond
M117 Done!             ; Update the LCD screen with "Done!"
```

With great power comes great responsibility — especially when you add custom movements into your G-code! Be sure to test your modifications from within your slicing software to ensure that you are not creating movements outside of your printer's boundaries. You can learn more about G-code for RepRap firmware and 3D printing at reprap.org/wiki/G-code. ⊘

COMMON *START.GCODE* AND *END.GCODE* COMMANDS

G COMMANDS

G1	Perform a synchronized movement
G21	Set all units to millimeters since 3D printers use the metric system
G28	Home the 3D printer or move the toolhead to the origin
G29	Use a probe to measure the flatness of the bed then compensate by "leveling" or "tramming" the bed via a live z offset
G90	Set all future commands to use absolute coordinates (as opposed to a relative position from the last location)
G92	Define the current physical position to user-specified values

M COMMANDS

M83	Set all future commands for the extruder to use relative coordinates from the last physical position (as opposed to absolute coordinates)
M84	Stop holding the current position of the motor
M104	Set the extruder temperature to a user-specified target (in Celsius) and immediately return control to the controller
M109	Set the extruder temperature to a user-specified target (in Celsius) and wait for the user-specified target to be achieved
M117	Display a user-specified message to appear on the LCD screen of the 3D printer
M140	Set the heated bed temperature to a user-specified target (in Celsius) and immediately return control to the controller
M190	Set the heated bed temperature to a user-specified target (in Celsius) and wait for the user-specified target to be achieved
M300	Play a beep sound based on a user-specified frequency and duration

Skill Builder

Kerf Wars

How much material is really removed by your cutter?

Written and photographed by Matt Stultz

MATT STULTZ is *Make:*'s digital fabrication editor, leader of our testing team, and the founder of 3DPPVD, Ocean State Maker Mill, and HackPGH.

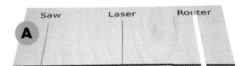

WE'VE ALL BEEN THERE — YOU START A WOODWORKING PROJECT AND measure all your boards, cut them all, start to assemble it, and for some reason everything is ¼" too short. "How did I screw up measuring everything by a quarter-inch? I must not be good at this!" Nope, you just didn't account for *kerf.*

Kerf is the material that a cutting tool like a saw or a router bit removes from your material (Figure **A**). We all grew up using scissors or shears to cut things — these shear the material, resulting in no material lost. But many cutting tools remove material. Think about it: saws make sawdust but scissors don't make paper dust.

So how can you ensure you leave enough room for kerf?

SAW — With saws it's easy. For starters, don't measure all your pieces at one time. Measure, then cut, then measure, and so on. Don't cut on your line; cut to the far side of it. The kerf will be removed from the unmeasured section and your board will be the correct length. Once you get good at it, you can measure the width of your saw blade's teeth (some teeth angle outward so they're wider than the body of the blade) and add that width into your board lengths, knowing it will be removed.

With digital fabrication tools, the best way to handle kerf is to plan for it in your design. In general, make holes smaller and joint spacing tighter to take up room for your kerf. But how much?

CNC ROUTER — Also easy. What's the width of your end mill: ⅛", ¼", ½"? The width of your bit will be the kerf removed from your part — on a good tool. If you really need tight tolerances, it's worth measuring your kerf in your chosen material to check for *runout*. Runout is when a spinning tool wobbles and doesn't stay centered; this can cause your kerf to be wider than expected. A bent end mill, bad collet or bearing, or loose spindle could all cause runout.

LASER CUTTER — Probably the hardest kerf to account for. You've seen those press-fit wooden boxes made on laser cutters; they make great project enclosures. But if the finger joints are too loose, they need glue to stay together; looser than that and it's hard to even square them up for gluing. Too tight and you'll never be able to press them together. You need the right kerf settings to get them to fit well.

With lasers, two major factors will decide how wide your kerf is:

» **Beam width** — This is established by the focal length set by the lens, but it's tricky. Beam width can change depending on the thickness of your material, so you might have more kerf on a ¼" sheet than an ⅛" sheet.

» **Material** — Wood gets burnt away by the laser, leaving behind nothing but the cut edge. But plastics may shrink away as they are not only cut but melted. This will make your kerf wider than expected.

Measuring the actual kerf cut by your laser is the way to know for sure. But this is tricky too — it's so narrow, it's hard to get a tool in to measure.

My trick to is to use a jig that I cut on a spare piece of material. I cut a *key* that's a given dimension (knowing that when it's cut, it will be smaller because of the kerf) and a series of *slots* to fit it into, each one 0.1mm smaller than the last. For instance, a 20mm key with 20, 19.9, 19.8, 19.7, 19.6, 19.5, and 19.4mm slots. Then I press the key into each slot and decide which fits best (Figure **B**).The actual kerf of my laser is half the difference between the key and the slot (because the kerf was removed from both sides). Download my jig at makezine.com/go/kerf-test-jig and try it yourself!

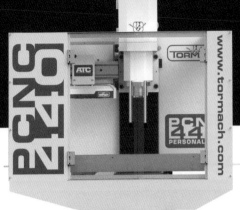

SHOW&TELL

Get inspired by some of our favorite submissions to Maker Share

If you'd like to see your project in a future issue of *Make:* magazine submit your work to makershare.com/missions/mission-make!

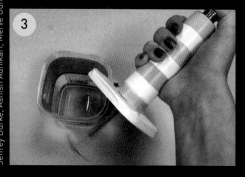

1 Woodworker **Jeffrey Burke** built this low profile hammock stand to take advantage of a gorgeous mountaintop vista. Made from reclaimed barn wood, it can be quickly taken apart and reassembled to get that perfect relaxing view. makershare.com/projects/reclaimed-wood-hammock-stand

2 **Ashish Adhikari** made this simple number plaque for his house using scrap wood, artificial turf, cardboard, and spray paint. Adhikari employed materials he had on hand, so the whole thing cost a whopping $0. makershare.com/projects/modern-house-number-plaque

3 Hailing from Istanbul, Turkey, **Merve Güngör** uses all kinds of everyday objects to make interesting DIY tools and creations. This mini mixer is hacked together using a candy container for the bowl, a simple motor, and a piece of aluminum can for the blade. Time to mix up some yummy treats! makershare.com/projects/lets-make-mini-mixer-recyclable-materials

4 In order to liven up his garden after dark, **Alan MacKay** built this light from a flickering LED and some bits and pieces around his shop. The specialty flickering LED package inspired the build, and a simple solar charging circuit keeps the faux-flame lit when it gets dark out. makershare.com/projects/create-flickering-candle-solar-garden-light

[+] Read about our Editors' Choice, Easy Soda Can Robot, on page 68.

Jeffrey Burke, Ashish Adhikari, Merve Güngör, Alan MacKay